Highlights of the Chinese Exposure Factors Handbook (Adults)

Highlights of the Chinese Exposure Factors Handbook (Adults)

Xiaoli Duan

Xiuge Zhao

Beibei Wang

Yiting Chen

Suzhen Cao

 Science Press
Beijing

 AMSTERDAM • BOSTON • HEIDELBERG • LONDON
NEW YORK • OXFORD • PARIS • SAN DIEGO
SAN FRANCISCO • SINGAPORE • SYDNEY • TOKYO
Academic Press is an imprint of Elsevier

Funded by

Ministry of Environmental Protection of P. R. China

Implemented by

Chinese Research Academy of Environmental Sciences
(CRAES)

National Center for Chronic and Noncommunicable
Disease Control and Prevention (NCNCD), Chinese
Center for Disease Control and Prevention (CDC)

Renmin University of China (RUC)

University of Science and Technology Beijing (USTB)

Authors

Highlights of the Chinese Exposure Factors Handbook (Adults)

Xiaoli Duan, Xiuge Zhao, Beibei Wang, Yiting Chen, Suzhen Cao

Contributors (In alphabetical order)
Chinese Research Academy of Environmental Sciences (CRAES)

Nan Huang, Jin Ma, Jing Nie, Yan Qian, Feifei Wang, Haiyan Wang, Hong Wang, Hongmei Wang, Xianliang Wang, Zongshuang Wang, Yongjie Wei, Zengguang Yan, Lixin Yang, Wenjie Zhang, Chanjuan Zheng

National Center for Chronic and Noncommunicable Disease Control and Prevention (NCNCD), Chinese Center for Disease Control and Prevention (CDC)

Zhengjing Huang, Nan Hu, Yong Jiang, Xiaoyan Li, Yichong Li, Yunna Sun, Limin Wang, Zhihui Wang, Xiangjun Yin, Mei Zhang, Yinjun Zhao

Renmin University of China (RUC)

Jing Guo, Xiaojuan Huang, Xiubin Wang

University of Science and Technology Beijing (USTB)

Ting Dong, Delong Fan, Tianxin Li, Xiang Liu, Yeqing Wang

Advisory Committee
(In alphabetical order)

Zhipeng Bai/*Chinese Research Academy of Environmental Sciences*

Bingheng Chen/*Fudan University*

Yude Chen/*Peking University Health Science Center*

Hongguang Cheng/*Beijing Normal University*

Michael Dellarco/*National Institute of Child Health and Human Development, USA*

Xinbiao Guo/*Peking University Health Science Center*

Jae-Yeon Jang/*Ajou University School of Medicine, Korea*

Haidong Kan/*Fudan University*

Fasheng Li/*Chinese Research Academy of Environmental Sciences*

Chunye Lin/*Beijing Normal University*

Wei Meng/*Chinese Research Academy of Environmental Sciences*

Jacqueline Moya/*Environmental Protection Agency, USA*

Xiaochuan Pan/*Peking University Health Science Center*

Yingwa Shen/*Chinese Research Academy of Environmental Sciences*

Dongchun Shin/*Yonsei University College of Medicine, Korea*

Kirk R. Smith/*University of California, Berkeley, USA*

Yonghui Song/*Chinese Research Academy of Environmental Sciences*

Chengye Sun/*Chinese Center for Disease Control and Prevention*

Shu Tao/*Peking University*

Jiansheng Wang/*Policy Research Center for Environment and Economy of Ministry of Environmental Protection*

Jinnan Wang/*Chinese Academy for Environmental Planning*

Linhong Wang/*Chinese Center for Disease Control and Prevention*

Wuyi Wang/*Chinese Academy of Sciences*

Fusheng Wei/*China National Environmental Monitoring Center*

Fengchang Wu/*Chinese Research Academy of Environmental Sciences*

Xuefang Wu/*Chinese Research Academy of Environmental Sciences*

Dongqun Xu/*Chinese Center for Disease Control and Prevention*

Qiujin Xu/*Chinese Research Academy of Environmental Sciences*

Zhencheng Xu/*South China Institute of Environmental Sciences, MEP*

Gonghuan Yang/*Peking Union Medical College*

Yunjiang Yu/*South China Institute of Environmental Sciences, MEP*

Jinliang Zhang/*Chinese Research Academy of Environmental Sciences*

Junfeng (Jim) Zhang/*Duke University, USA*

Binghui Zheng/*Chinese Research Academy of Environmental Sciences*

Yun Zhou/*Chinese Research Academy of Environmental Sciences*

Preface

Exposure assessment is the central component of environmental health sciences. It lies in the middle of the causal pathway leading from the sources of pollution at one end to the health effects of concern at the other. It thus is the best single indicator both of the potential risk of the pollutant and of the efficacy of alternative ways of controlling the pollution. Ideally, of course, exposure assessment is done by measurements in the population of concern doing their normal daily activities that bring them into contact with the pollutants. In practice, however, it is not possible to conduct individual assessments for every situation or to model changes in exposure in detail based on future scenarios.

An exposure factor handbook such as this provides a convenient and practical solution to obtaining reasonable estimates of exposure to important pollutants in various populations. It provides estimates of the exposure in different settings that have been derived from past studies linked to standard metrics, such as emissions. It can be used to scope the potential extent of the exposure and health impacts of a wide variety of situations quickly and at low cost to evaluate whether more detailed assessments and modeling might be warranted. It serves a range of users, from regulators, industry environmental and occupational control departments, researchers, and citizen groups.

Although there are similarities across populations, exposure factors vary throughout the world according to local circumstances. Given the growing concern with the health implications of environmental pollution in China, it is very appropriate, therefore, that this Chinese exposure factor handbook be published now. It will serve many useful purposes, although of course will need to be revisited as conditions change and knowledge improves.

It takes much dedication and perseverance to collect all the many types of data needed and to put them into a useful form for this kind of handbook. Dr. Duan and her many colleagues are to be highly commended for accomplishing this major effort for the first time in China.

Kirk R. Smith
Professor of Global Environmental Health
School of Public Health,
University of California, Berkeley
Member of US National Academy of Sciences
Chair, Exposure Science in the 21st Century: A Vision and a Strategy
US National Research Council Report, 2012
July, 2014

Foreword

In the last three decades, China has achieved rapid economic growth, which has brought hundreds of millions of people out of poverty. However, at the same time the country is facing increasingly severe environmental challenges, which threaten not only sustainable economic and social development, but also human health. Policies of "modifying industrial structure" and "scientific development" have been proposed by the central government in recent years, to balance economic development and environmental protection. Environmental health risk assessment (EHRA), an essential tool for scientific decision-making, has a vital role in the process of environmental management.

Exposure factors are basic parameters for assessing the potential risk of exposure that are based on the physiological and behavioral characteristics of humans. With regard to *The 12th five-year plan for the environmental health work of national environmental protection* (MEP, 2011), the Chinese Research Academy of Environmental Sciences (CRAES), entrusted by the Department of Science, Technology and Standards of China's Ministry of Environmental Protection (MEP), had conducted the *Environmental exposure related activity pattern research for the Chinese population (Adults)* (CEERHAPS-A) from 2011 to 2012. Based on this survey, the *Exposure factors handbook of Chinese population (Adults)* (CEFH-A) and *Report of environmental exposure related activity patterns for the Chinese population* were compiled, and published in 2013. The *Highlights of the Chinese exposure factors handbook (Adults)* is a brief introduction to the content of the CEFH-A, designed to provide a reference for assessors, scientists, and managers, in which exposure factors such as inhalation rates, water ingestion rates, food intake, time-activity related to exposure, body weight, surface area, life expectancy, and residential factors are discussed.

CRAES was responsible for the preparation of the CEFH-A, and the *Highlights of the Chinese exposure factors handbook (Adults)*. Dr. Xiaoli Duan served as the principal investigator and the chief editor for both books, providing overall direction and assistance on the organization and execution of the work. The Chinese Centers for Disease Control and Prevention (CDC) made important contributions to the site studies and data collection. The advisory committee consisted of more than 30 experts from 17 academic institutes, offering technical support, scientific advice, and critical comments throughout the entire project.

Since the CEFH-A is the first exposure factors handbook for Chinese population, there are some unavoidable deficiencies due to the limited time and experience available. Continuous revisions and data updates will be required in the future, and data for soil ingestion, air exchange rates, and the exposure of special populations are also required. More details are available in the *Exposure Factors Handbook of Chinese Population (Adults)*. Your suggestions and comments would be greatly appreciated.

President of Chinese Research Academy of Environmental Sciences,

Member of Chinese Academy of Engineering

June, 2014

Synopsis

The *Highlights of the Chinese exposure factors handbook (Adults)* is a brief introduction to the content of the *Exposure factors handbook for the Chinese population (Adults)* (CEFH-A), which is designed to provide a reference for assessors, scientists, and decision makers, who concern about environment and health. The handbook considers exposure factors such as inhalation rates, water ingestion rates, food intake, time-activity related to exposure, body weight, surface area, life expectancy, and residential factors.

In each chapter, definitions, possible influencing factors, and the survey methods used to determine the factors are introduced. Information is given for urban/rural location, gender, age group, and region, with recommended values provided for the mean, median and 5th, 25th, 75th, and 95th percentile values.

Contents

List of Tables

1 INTRODUCTION

1.1 Background and purpose

Exposure factors represent human behavior and characteristics, and include intake/ingestion factors, time-activity factors, physical (body weight, skin surface area, inhalation rate, life expectancy) and other environment-related (floor area, ventilation rate, heating duration) factors. The values of these factors are different in various countries/regions due to the diversity of social environments, economic conditions, race, climates, and geographic locations. Exposure factors are often used to determine the exposure risks of individuals or a target population, and play an important role in the process of risk-based environmental management or policy making.

The United States Environmental Protection Agency (USEPA) first developed the *Exposure factors handbook* in 1989, and revised it in 1997 and 2011 (Phillips and Moya, 2013). By integrating the existing data of 30 member states, the European Union (EU) established an exposure factors database (ExpoFacts) and tool kit, which was made available on the Internet in 2007 (ECJRC, 2012). The *Japanese exposure factors handbook* (AIST, 2007) was compiled by the National Institute of Advanced Industrial Science and Technology in 2007. In the same year, the *Korea exposure factors handbook* was issued by the Ministry of the Environment of Korea (Jang et al., 2007).

China has not previously published an exposure factors handbook. However, with its rapid economic development, the largest population in the world, and concerns about environmental quality, China is faced with the need for exposure assessment to prevent environmental health risks.

China's MEP started the development of *Exposure factors handbook for the Chinese population* (CEFH) program from 2011 based on *The 12th five-year plan for the environmental health work of national environmental protection* (MEP, 2011). In this program, the *Exposure factors handbook for the Chinese population (Adults)* (CEFH-A) (≥18 years old) (MEP, 2013a) and the *Exposure factors handbook for the Chinese population (Children)* (CEFH-C) (<18 years old) were developed separately to ensure better representativeness of the two groups. The compiling of the CEFH-A occurred during the first stage from 2011 to 2012, and the production of the CEFH-C was scheduled for the second stage from 2013 to 2015.

At the end of 2013, the *Exposure factors handbook for the Chinese population (Adults)* (CEFH-A) (MEP, 2013a) and the *Report of environmental exposure related activity patterns research for the Chinese population (Adults)* (MEP, 2013b) were completed and published.

The *Highlights of the Chinese exposure factors handbook (Adults)* is a brief introduction to the content of the Handbook.

1.2 Targeted exposure factors

Exposure factors discussed in the highlights and handbook are classified into four categories.

(1) Inhalation factors

Inhalation rates (Chapter 2);

(2) Ingestion factors

Water ingestion rates (Chapter 3);

Food intake (Chapter 4).

(3) Time-activity factors

Time-activity related to air exposure (Chapter 5);

Time-activity related to water exposure (Chapter 6);

Time-activity related to soil exposure (Chapter 7);

Time-activity related to electromagnetic exposure (Chapter 8).

(4) Other exposure factors

Body weight (Chapter 9);

Surface area (Chapter 10);

Life expectancy (Chapter 11);

Residential factors (Chapter 12).

Recommendations for the Chinese population for each factor are summarized in Table 1-1, and detailed information, such as the recommended values for different groups and the distribution of data, are presented in the relevant chapters. A comparison of recommendations between China and other countries is given in Table 1-2 to provide an overview of exposure factor values.

Table 1-1 Recommended values of exposure factors for the Chinese population

Category		Total	Gender		Age				Urban/Rural[1]	
			Male	Female	18-44	45-59	60-79	80+	Urban	Rural
Long-term inhalation rate (m³/d)	Sleeping	15.7	18.0	14.5	16.0	16.0	13.7	12.0	15.8	15.6
Short-term inhalation rate (L/min)	Sedentary behavior	5.5	6.3	5.1	5.8	5.7	4.7	4.1	5.5	5.5
	Light-intensity activity	6.6	7.5	6.1	6.9	6.9	5.7	5.0	6.6	6.5
	Moderate-intensity activity	8.2	9.4	7.6	8.6	8.6	7.1	6.2	8.3	8.2
	High-intensity activity	21.9	25.1	20.3	23.0	22.9	19.0	16.5	22.1	21.8
	Very-high-intensity activity	32.9	37.7	30.4	34.5	34.4	28.5	24.8	33.1	32.7
Drinking water intake (ml/d)[2]	Total drinking water intake	54.8	62.8	50.7	57.6	57.3	47.4	41.4	55.1	54.6
	Total drinking water intake	1,850	2,000	1,713	1,875	1,900	1,800	1,525	1,900	1,825
	Direct drinking water intake	1,125	1,250	1,000	1,125	1,125	1,000	875	1,250	1,100
	Indirect drinking water intake	480	500	450	450	500	500	450	400	600
Food intake (g/d)[*3]	Total food intake	1,056.6	—	—	—	—	—	—	1,117.7	1,033.0
	Rice products	238.3	—	—	—	—	—	—	217.8	246.2
	Wheat products	140.2	—	—	—	—	—	—	131.9	143.5
	Other grain products	23.6	—	—	—	—	—	—	16.3	26.4
	Potato products	49.1	—	—	—	—	—	—	31.9	55.7
	Dark color vegetables[4]	90.8	—	—	—	—	—	—	88.1	91.8
	Light color vegetables[4]	185.4	—	—	—	—	—	—	163.8	193.8
	Fruits	45.0	—	—	—	—	—	—	69.4	35.6
	Pork products	50.8	—	—	—	—	—	—	60.3	47.2
	Livestock meat products[5]	9.2	—	—	—	—	—	—	15.5	6.8
	Poultry products	13.9	—	—	—	—	—	—	22.6	10.6
	Fish and shellfish	29.6	—	—	—	—	—	—	44.9	23.7
	Dairy products	26.5	—	—	—	—	—	—	65.8	11.4
	Egg products	23.7	—	—	—	—	—	—	33.2	20.0

Continued

Category		Total	Gender		Age				Urban/Rural [1]	
			Male	Female	18-44	45-59	60-79	80+	Urban	Rural
Activity patterns	Time outdoors (min/d)	221	236	209	219	235	210	150	180	255
	Time indoors (min/d)	1,200	1,185	1,215	1,201	1,185	1,203	1,260	1,239	1,165
	Time showering/bathing (min/d)	7	7	7	8	7	6	5	8	6
	Time in transit (min/d)	45	50	40	50	50	40	30	50	40
	Proportion of population that swim (%)	3.3	5.3	1.5	5.5	2.3	1.0	0.3	4.0	2.6
	Time swimming [6] (min/month)	155	154	159	148	181	169	117	180	123
	Proportion of population in contact with soil [7] (%)	47.1	48.5	46.0	44.4	53.0	44.7	18.2	21.6	68.7
	Time in contact with soil [6,7] (min/d)	204	212	195	205	215	183	112	168	214
	Proportion of population using a cell phone (%)	76.4	79.9	73.6	89.1	78.5	54.3	23.7	83.2	70.6
	Time talking on a cell phone [6] (min/d)	24	26	22	28	20	14	12	28	21
	Proportion of population using a computer (%)	29.5	32.1	27.3	47.2	22.8	10.1	4.6	43.2	17.9
	Time using computer [6] (min/d)	167	162	173	175	144	138	146	188	134
Body weight (kg)	Body	60.6	65.0	56.8	60.1	62.4	59.4	54.3	62.0	59.7
Skin-surface area (m²)	Body	1.6	1.7	1.5	1.6	1.6	1.6	1.5	1.6	1.6
	Head	0.12	0.13	0.12	0.12	0.12	0.12	0.11	0.12	0.12
	Trunk	0.60	0.63	0.57	0.61	0.61	0.59	0.56	0.61	0.60
	Arms	0.24	0.25	0.23	0.24	0.24	0.23	0.22	0.24	0.24
	Hands	0.08	0.08	0.07	0.08	0.08	0.08	0.07	0.08	0.08
	Legs	0.47	0.49	0.44	0.47	0.47	0.46	0.43	0.47	0.46
	Feet	0.11	0.11	0.10	0.11	0.11	0.10	0.10	0.11	0.10

Continued

Category		Total	Gender		Age				Urban/Rural [1]	
			Male	Female	18-44	45-59	60-79	80+	Urban	Rural
Life expectancy *[8] (years)		74.83	72.38	77.37	—	—	—	—	—	—
Building Parameters	Floor area [9] (m²)	100	—	—	—	—	—	—	92	106
	Heating duration [6,10] (d/year)	110	—	—	—	—	—	—	120	105
	Ventilation rate [11] (min/d)	465	—	—	—	—	—	—	465	453

Source: 2013 CEERHAPS-A, except data for food intake and life expectancy.

* The data are mean values, whereas all other data presented above are median values.

1. The division of urban and rural was based on the CNBS regulations, in which the urban area includes cities and townships.

2. Milk, beverages, wine, and water in purchased food materials were not included.

3. Source: 2002 China National Nutrition and Health Survey. Results presented as the daily intake of a male adult (aged 18 years) participating in light physical activity.

4. Dark color vegetables include dark green vegetables, such as spinach, edible rape, celery leaf, red or orange vegetables (e.g., tomatoes, carrots, and pumpkin), and purple vegetables (e.g., red amaranth and purple cabbage). Light color vegetables include cabbage, celery cabbage, white radish, etc.

5. Includes beef, mutton, ass meat, horse meat, rabbit meat, and food comprising these meats.

6. Doers only.

7. Includes soil and outdoor settled dust.

8. Source: 2012 China Statistical Yearbook, using life expectancy at birth for 2010.

9. The floor area excludes open spaces, such as a balconies and courtyards, as well as rarely used indoor areas (e.g., storerooms and basements).

10. Accumulated heating time in one year.

11. The longest average ventilation time among the bedroom, study, and living room was reported as the ventilation rate for the household.

Table 1-2 Comparison of recommended values for the Chinese population with other international sources

Exposure factors	Category	China	Asian Japan[1]	Asian Korea[2]	Oceania Australia[3]	North America USA[4]	North America Canada[5]	Europe EU[6]
Long-term inhalation rates (m³/d)	Male	18.0*	17.3	15.7	15	14.7	18.0	—
	Female	14.5*	—	12.8	—	—	15.3	—
Short-term inhalation rates (L/min)								
Sleeping	Combined	5.5*	—	—	—	4.3-5.3	—	—
Sedentary behavior	Combined	6.6*	—	7.50	—	4.2-5.3	—	—
Light-intensity activity	Combined	8.2*	—	15.67	—	12-13	—	—
Moderate-intensity activity	Combined	21.9*	—	18.83	—	25-29	—	—
High-intensity activity	Combined	32.9*	—	31.17	—	47-53	—	—
Very-high intensity activity	Combined	54.8*	—	38.83	—	—	—	—
Direct drinking water intake (ml/d)	Male	1,250*	668[a]	1,660	2,000[b]	1,043	—	375-971
	Female	1,000*	666[a]	1,346	—	—	—	—
Total food intake (g/d)	Male	1,056.6	1,224.3[c]	—	1,550	2,494[d]	—	—
	Female		1,099.1[c]	—	1,200	2,117[d]	—	—
Time-activity patterns								
Time outdoors (h/d)	Combined	3.68*	1.20	1.27	3	4.8	1.02	1
Time indoors (h/d)	Combined	20*	—	21.35	20	19	22.98	21
Time in vehicle/transit (h/d)	Combined	0.75*[e]	—	1.38	1	1.6	—	2
Time showering/bathing (min/d)	Combined	7*	S: 6.6[f] B: 25.2[f]	S: 16.8[i] B: 3.3[i]	S:8 B:21	S:12.5 B:17.5	37[h]	—
Time swimming (min/mon)	Combined	155[i]	268.5[i]	292.2[i]	900[i,j]	45*	—	—
Proportion of population that swims (%)	Combined	3.3	7.5	7.0	—	—	—	—
Body weight (kg)	Male	65.0	64.0	69.2	85	86	83.3[k]	80
	Female	56.8	52.7	56.4	70	73	69.8[k]	67

Continued

Exposure factors	Category	China	Asian		Oceania	North America		Europe
			Japan [1]	Korea [2]	Australia [3]	USA [4]	Canada [5]	EU [6]
Total surface area (cm²)	Male	17,000*	16,900	18,318	21,000	20,700	20,182	20,300
	Female	15,000*	15,100	15,853	19,000	18,300	17,784	17,700
Life expectancy [l] (years)	Male	72.38	77.72	75.1	79	75	80.9	64.2-77.3
	Female	77.37	84.60	81.9	84	80		73.5-83.1
Floor area (m²)	Combined	100*	92.50	—	180	201	—	33.8-114.9

* Median values.
1. Source: Japanese exposure factors handbook (AIST Research Center for CRM, 2007).
2. Source: Korean exposure factors handbook (Jang et al., 2007).
3. Source: Australian exposure factor guide (Office of Health Protection of the Australian Government Department of Health, 2012).
4. Source: Exposure factors handbook: 2011 Edition (USEPA, 2011).
5. Source: Canadian exposure factors handbook (GM Richardson and Stantec Consulting Ltd., 2013).
6. Source: Expofacts database (ECJRC, 2012), (Phillips and Moya, 2014).
a. Not including water ingestion through the drinking of tea.
b. Life expectancy water intake. This value is used in the evaluation of long-term exposure.
c. The value is the sum of the recommendations for all foods.
d. The value was converted following the unification of units. The consumption of beverages and water ingestion is included
e. Mean value of the total time spent in transportation.
f. Average value of male and female.
g. The value was calculated by recommendations of duration and frequency.
h. Including dressing time.
i. Doers only.
j. The value was converted following the unification of units. Assumed that all outdoor activity was swimming.
k. Average body weight of adults (20 to 65 years old).
l. Using life expectancy at birth.

1.3 The approach to developing recommended values of exposure factors

1.3.1 Literature searches

Before conducting the literature search, the priorities for all of the exposure factors were evaluated based on the *Risk assessment guidance for superfund* (USEPA, 1989). Data from reports and literatures with open sources were collected, and the reliability, suitability, and availability of data were evaluated (Table 1-3).

Table 1-3 Criteria for the evaluation of research data

General considerations		Criteria
Reliability	Research method	The latest research method or technologies were used
	Sample size	Priority given to the research with larger sample sizes
	Response rate	Acceptance criteria of response rate was greater than 80% (face to face interviews) or 70% (telephone or e-mail interviews)
	Quality assurance	QA/QC measures were applied during the survey
	Research design	Good sampling design was employed to minimize bias
	Uncertainty	Uncertainty was described in the study
	Originality of data	Priority was given to original data
Applicability	Study purpose	Priority was given to studies designed to obtain exposure factors
	Representative of population	Studies targeted to the Chinese population, with detailed information of subject classification (by gender and age group) and data distribution
	Time information	The latest study was chosen to represent the current situation
Availability	Data availability	Data was either open source or publicized

However, during the evaluation, data gaps were discovered due to small sample sizes or the inadequacy of grouped data. Therefore, the *Environmental exposure related activity pattern research for the Chinese population (Adults)* (CEERHAPS-A) was conducted from 2011 to 2012 to collect data.

1.4 Uncertainty and variability

2013 CEERHAPS-A is the primary data source of the hand-book for the recommendations of inhalation rates, water inges-tion rates, time-activity factors, body weight and residential fac-tors are derived from analysis of the survey result. Therefore, the variability and uncertainty of 2013 CEERHAPS-A are discussed to evaluate the confidence in recommendation.

1.4.1 Uncertainty

Poor data quality is an important source of the uncertainty for recommendations, which can be induced by biased sample design, limited sample size, or error from measurement.

In 2013 CEERHAPS-A, a multi-stage cluster random sampling method was used, with demographic characteristics, rural/urban status and gender as stratified factors, and a total of 91,527 samples were collected, which covered 31 provinces in mainland China. Face to face interviews were conducted to get biological information and time-activity patterns of the interviewees. All of the surveyors were trained to ensure that they had the same understanding of each question. Body weight and height were measured by uniform measuring tools with the application of QA/QC procedures. After data screening, 91,121 questionnaires were administered to determine exposure factors, with an overall response rate of 94.6%. Table 1-4 gives the distribution of interviewees by selected factors.

Table 1-4 Distribution of interviewees by selected factors

Category		Sample size	Proportion (%)
Total		91,121	100.0
Urban/Rural	Urban	41,826	45.9
	Rural	49,295	54.1
Gender	Male	41,296	45.3
	Female	49,825	54.7
Age	18-44	36,682	40.3
	45-59	32,374	35.5
	60-79	20,579	22.6
	80+	1,486	1.6
Region	North	18,097	19.9
	East	22,965	25.2
	South	15,184	16.7
	Northwest	11,271	12.4
	Northeast	10,179	11.2
	Southwest	13,425	14.7

Chi-square test (Table 1-5) shows there was no statistically significant difference between age and gender proportion of respondents in 2013 CEERHAPS-A and China's sixth nationwide census.

Table 1-5 Comparison check of weighted sample by Chi-square test

Age	Total Sample (%)	Chinese census (%)	χ^2	Male Sample (%)	Chinese census (%)	χ^2	Female Sample (%)	Chinese census (%)	χ^2
18-	5.40	16.10	7.11	6.03	16.12	6.32	4.88	16.09	7.81
25-	5.78	9.59	1.51	6.07	9.55	1.27	5.54	9.62	1.73
30-	6.94	9.22	0.56	6.95	9.30	0.59	6.93	9.13	0.53
35-	9.21	11.20	0.35	8.99	11.34	0.49	9.39	11.05	0.25
40-	12.93	11.84	0.10	12.57	11.95	0.03	13.23	11.73	0.19
45-	13.55	10.02	1.24	12.71	10.10	0.67	14.25	9.94	1.87
50-	10.16	7.47	0.97	9.82	7.58	0.66	10.45	7.36	1.29
55-	11.81	7.72	2.17	11.60	7.72	1.96	11.99	7.72	2.37
60-	9.40	5.57	2.64	9.66	5.60	2.94	9.18	5.53	2.41
65-	6.05	3.90	1.18	6.38	3.90	1.58	5.78	3.91	0.90
70-	4.29	3.13	0.43	4.54	3.08	0.69	4.08	3.18	0.26
75-	2.85	2.26	0.15	3.05	2.12	0.41	2.68	2.41	0.03

Continued

Age	Total			Male			Female		
	Sample (%)	Chinese census (%)	χ^2	Sample (%)	Chinese census (%)	χ^2	Sample (%)	Chinese census (%)	χ^2
80+	1.63	1.99	0.06	1.63	1.65	0	1.63	2.34	0.21
	$\chi^2=18.49, P > 0.05$			$\chi^2=17.61, P > 0.05$			$\chi^2=19.86, P > 0.05$		

Therefore, the recommendations derived from 2013 CEER-HAPS-A have a low uncertainty.

1.4.2 Variability

The variability always exists, since there are variations in spatial and temporal distribution, within an individual or among the individuals (USEPA, 2011). In this handbook, the variability is addressed by disaggregating the population with the same characteristic and presenting data distribution information of each factor.

Gender and age (18-44, 45-59, 60-79 and 80+) are considered as the necessary taxonomy for the exposure assessment due to the variation of life style and susceptibility to toxic (Committee on Improving Risk Analysis Approaches, 2011). And in China, urban/rural population or people in various locations (e.g., southwest versus east) have different economic conditions, dietary habits and life styles, thus urban/rural location and region are other classifications that cannot be neglected. Table 1-6 lists the classification of provinces by region based on the geographical distribution and the analysis of socioeconomic state.

Table 1-6 Classification of provinces by region

Region	Province/Municipality/Autonomous region
North	Beijing, Tianjin, Hebei, Shanxi, Neimenggu, Henan
East	Shanghai, Jiangsu, Zhejiang, Fujian, Shandong, Anhui, Jiangxi
South	Hubei, Hunan, Guangdong, Guangxi, Hainan
Northwest	Shaanxi, Gansu, Qinghai, Ningxia, Xinjiang
Northeast	Jilin, Heilongjiang, Liaoning
Southwest	Chongqing, Sichuan, Guizhou, Yunnan, Xizang

The mean value and percentile values (P5, P25, P50, P75 and P95) are summarized to present the variability of each factor, and with this information, uncertainty and variability are suggested to be simulated by models like Monte Carlo analysis.

1.5 Limitation and further research needs

(1) The average or median data for province is listed for reference. However, the results of 2013 CEERHAPS-A is technically not available to represent the entire population in each province due to the limited sample size. Thus, more data is needed to improve the provincial representativeness.

(2) Some factors like air exchange rates or soil/dust ingestion for the Chinese population are limited, although soil/dust ingestion rates are discussed in Chapter 5 of the handbook, with the data currently based on the recommendation of the American rates (USEPA, 2011). Further research is needed.

(3) Research is needed for exposure factors of special populations, like children and people occupied in high risk jobs.

(4) The value of the exposure factors may change with different behaviors and activity habits, as well as with the improvements in experimental technologies and analytic methods. Therefore, periodic updates are necessary.

(Xiaoli Duan, Yiting Chen, Limin Wang, Yong Jiang,

Tianxin Li, Jing Guo)

2 INHALATION RATES

2.1 Introduction

Inhalation rates refer to the volume of air inhaled per unit time (GM Richardson and Stantec Consulting Ltd., 2013). The recommendations include long and short-term inhalation rates. Based on the activity level, short-term inhalation rates could be further divided into inhalation rates for sleeping, sedentary behavior, and light, moderate, high, and very high intensity activity.

There are three approaches to obtaining inhalation rates.

(1) Doubly labeled water measurements

Two forms of stable isotopically labeled water (^2H_2O and $H_2^{18}O$) are used, and the difference in the disappearance rates of the two isotopes in a subject's body represents the energy expended over a period of 1-3 half-lives of the labeled water.

(2) Relationship between inhalation rates and heart rates

This approach measures the inhalation rates and heart rates

of a representative population, and establishes a model of the relationship between them. The inhalation rates of the target population can then be calculated based on the model (USEPA, 2011).

(3) Human energy expenditure method

$$IR = \frac{E \times H \times VQ}{1440} \qquad (2\text{-}1)$$

Where, IR: Inhalation rate, L/min; H: volume of oxygen consumed to produce 1 kcal of energy (typically 0.05 L/kJ); VQ: ventilatory equivalent (typically 27); E: energy expended per unit time, kJ/d.

2.2 Recommended values

The recommended inhalation rates for adults were calculated based on data collected in the *Environmental exposure related activity patterns research of Chinese population (Adults)* (CEERHAPS-A), using the energy expenditure method. Table 2-1 presents the mean and 5th-95th percentile data of long-term inhalation rates by gender, age, urban/rural location, and region. Tables 2-2 to 2-7 provide the mean and 5th-95th percentile data of short-term inhalation rates for individuals performing various activities by gender, age, urban/rural location, and region. Table 2-8 shows the recommended values of long and short-term inhalation rates by province.

Table 2-1 Recommended values of long-term inhalation rates

Category		Long-term inhalation rates (m³/d)					
		Mean	P5	P25	P50	P75	P95
Total		16.1	12.3	14.2	15.7	18.0	20.4
Gender	Male	17.7	12.7	16.7	18.0	19.2	21.3
	Female	14.5	12.0	13.8	14.5	15.3	16.5
Age	18-44	16.7	13.5	14.6	16.0	18.2	20.6
	45-59	16.7	13.8	14.8	16.0	18.1	20.2
	60-79	13.8	11.2	12.6	13.7	15.0	17.5
	80+	12.0	9.9	11.1	12.0	13.1	15.0
Urban/Rural	Urban	16.3	12.6	14.3	15.8	18.3	20.7
	Rural	16.0	12.1	14.1	15.6	17.8	20.1
Region	North	16.5	12.7	14.6	15.9	18.4	20.9
	East	16.1	12.2	14.2	15.6	18.1	20.4
	South	15.8	12.0	14.0	15.4	17.6	19.7
	Northwest	16.2	12.6	14.4	15.8	18.2	20.4
	Northeast	16.7	12.9	14.7	16.2	18.6	21.2
	Southwest	15.7	11.8	13.9	15.3	17.5	19.7

Source: 2013 CEERHAPS-A.

Table 2-2 Recommended values of short-term inhalation rates—sleeping

Category		Sleeping (L/min)					
		Mean	P5	P25	P50	P75	P95
Total		5.6	4.3	5.0	5.5	6.3	7.1
Gender	Male	6.2	4.5	5.8	6.3	6.7	7.4
	Female	5.1	4.2	4.8	5.1	5.3	5.8
Age	18-44	5.8	4.7	5.1	5.8	6.5	7.3
	45-59	5.8	4.8	5.2	5.7	6.4	7.1
	60-79	4.8	3.9	4.4	4.7	5.2	6.1
	80+	4.2	3.4	3.8	4.1	4.5	5.1
Urban/Rural	Urban	5.7	4.4	5.0	5.5	6.4	7.2
	Rural	5.6	4.2	4.9	5.5	6.2	7
Region	North	5.8	4.4	5.1	5.6	6.4	7.3
	East	5.6	4.3	5.0	5.5	6.3	7.1
	South	5.5	4.2	4.9	5.4	6.2	6.9
	Northwest	5.7	4.4	5.0	5.5	6.3	7.1
	Northeast	5.8	4.5	5.1	5.7	6.5	7.4
	Southwest	5.5	4.1	4.9	5.4	6.1	6.9

Source: 2013 CEERHAPS-A.

Table 2-3 Recommended values of short-term inhalation rates—sedentary behavior

Category		Sedentary behavior (L/min)					
		Mean	P5	P25	P50	P75	P95
Total		6.8	5.1	6.0	6.6	7.6	8.6
Gender	Male	7.4	5.3	7.0	7.5	8.1	8.9
	Female	6.1	5.0	5.8	6.1	6.4	6.9
Age	18-44	7.0	5.6	6.1	6.9	7.7	8.8
	45-59	7.0	5.8	6.3	6.9	7.7	8.5
	60-79	5.8	4.6	5.2	5.7	6.3	7.3
	80+	5.0	4.1	4.6	5.0	5.4	6.2
Urban/Rural	Urban	6.8	5.3	6.0	6.6	7.7	8.7
	Rural	6.7	5.1	5.9	6.5	7.5	8.4
Region	North	6.9	5.3	6.1	6.7	7.7	8.8
	East	6.8	5.1	6.0	6.5	7.6	8.6
	South	6.6	5.0	5.9	6.4	7.4	8.3
	Northwest	6.8	5.3	6.0	6.6	7.6	8.5
	Northeast	7.0	5.4	6.2	6.8	7.8	8.9
	Southwest	6.6	4.9	5.8	6.4	7.3	8.3

Source: 2013 CEERHAPS-A.

Table 2-4 Recommended values of short-term inhalation rates
—light-intensity activity

Category		Light-intensity activity (L/min)					
		Mean	P5	P25	P50	P75	P95
Total		8.4	6.4	7.5	8.2	9.5	10.7
Gender	Male	9.3	6.7	8.7	9.4	10.1	11.1
	Female	7.6	6.3	7.2	7.6	8.0	8.7
Age	18-44	8.7	7.0	7.7	8.6	9.7	10.9
	45-59	8.7	7.3	7.8	8.6	9.6	10.6
	60-79	7.2	5.8	6.5	7.1	7.8	9.2
	80+	6.3	5.2	5.8	6.2	6.7	7.7
Urban/Rural	Urban	8.5	6.6	7.5	8.3	9.6	10.8
	Rural	8.4	6.3	7.4	8.2	9.3	10.6
Region	North	8.6	6.7	7.7	8.3	9.6	11.0
	East	8.4	6.4	7.4	8.2	9.5	10.7
	South	8.3	6.3	7.3	8.1	9.2	10.3
	Northwest	8.5	6.6	7.5	8.3	9.5	10.7
	Northeast	8.7	6.8	7.7	8.5	9.8	11.1
	Southwest	8.2	6.2	7.3	8.0	9.2	10.3

Source: 2013 CEERHAPS-A.

**Table 2-5 Recommended values of short-term inhalation rates
—moderate-intensity activity**

Category		Moderate-intensity activity (L/min)					
		Mean	P5	P25	P50	P75	P95
Total		22.5	17.2	19.9	21.9	25.2	28.5
Gender	Male	24.8	17.8	23.3	25.1	26.9	29.7
	Female	20.2	16.8	19.2	20.3	21.3	23.1
Age	18-44	23.3	18.8	20.4	23.0	25.8	29.2
	45-59	23.3	19.4	20.8	22.9	25.6	28.4
	60-79	19.3	15.5	17.4	19.0	20.8	24.5
	80+	16.8	13.7	15.3	16.5	18.0	20.6
Urban/Rural	Urban	22.8	17.6	20.1	22.1	25.6	28.9
	Rural	22.3	16.9	19.8	21.8	24.9	28.2
Region	North	23.0	17.8	20.4	22.3	25.7	29.2
	East	22.5	17.1	19.9	21.8	25.3	28.6
	South	22.0	16.8	19.6	21.5	24.6	27.6
	Northwest	22.7	17.6	20.1	22.1	25.4	28.5
	Northeast	23.3	18.1	20.6	22.6	26.0	29.7
	Southwest	21.9	16.5	19.5	21.4	24.5	27.5

Source: 2013 CEERHAPS-A.

**Table 2-6 Recommended values of short-term inhalation rates
—high-intensity activity**

Category		High-intensity activity (L/min)					
		Mean	P5	P25	P50	P75	P95
Total		33.8	25.7	29.8	32.9	37.8	42.8
Gender	Male	37.2	26.7	35.0	37.7	40.3	44.6
	Female	30.3	25.2	28.9	30.4	32.0	34.7
Age	18-44	35.0	28.2	30.6	34.5	38.7	43.8
	45-59	35.0	29.1	31.3	34.4	38.4	42.5
	60-79	29.0	23.2	26.1	28.5	31.3	36.7
	80+	25.2	20.6	23.0	24.8	27.0	30.8
Urban/Rural	Urban	34.2	26.4	30.1	33.1	38.4	43.4
	Rural	33.5	25.3	29.7	32.7	37.4	42.2
Region	North	34.5	26.7	30.6	33.4	38.6	43.8
	East	33.8	25.6	29.8	32.7	38.0	42.9
	South	33.0	25.2	29.3	32.2	37.0	41.4
	Northwest	34.0	26.4	30.1	33.2	38.1	42.7
	Northeast	34.9	27.1	30.8	33.9	39.0	44.5
	Southwest	32.8	24.7	29.2	32.1	36.7	41.3

Source: 2013 CEERHAPS-A.

19

Table 2-7 Recommended values of short-term inhalation rates —very-high-intensity activity

Category		Very-high-intensity activity (L/min)					
		Mean	P5	P25	P50	P75	P95
Total		56.3	42.9	49.7	54.8	63.0	71.3
Gender	Male	62.0	44.5	58.3	62.8	67.1	74.3
	Female	50.5	41.9	48.1	50.7	53.3	57.8
Age	18-44	58.3	47.0	51.1	57.6	64.5	72.9
	45-59	58.3	48.4	52.1	57.3	63.9	70.9
	60-79	48.3	38.6	43.6	47.4	52.1	61.2
	80+	41.9	34.3	38.3	41.4	45.0	51.4
Urban/Rural	Urban	56.9	44.0	50.1	55.1	63.9	72.3
	Rural	55.8	42.2	49.4	54.6	62.3	70.4
Region	North	57.5	44.5	51.0	55.6	64.3	73.0
	East	56.3	42.7	49.6	54.6	63.3	71.4
	South	55.1	42.0	48.9	53.7	61.6	69.0
	Northwest	56.7	43.9	50.2	55.3	63.5	71.1
	Northeast	58.2	45.2	51.4	56.6	65.0	74.2
	Southwest	54.7	41.1	48.7	53.5	61.2	68.8

Source: 2013 CEERHAPS-A.

Table 2-8 Recommended values of inhalation rates by province *

Province/ Municipality/ Autonomous region	Long-term inhalation rates (m³/d)	Short-term inhalation rates (L/min)					
		Sleeping	Sedentary behavior	Light intensity activity	Moderate intensity activity	High intensity activity	Very high intensity activity
Total	15.7	5.5	6.6	8.2	21.9	32.9	54.8
Beijing	16.1	5.6	6.8	8.4	22.5	33.8	56.3
Tianjin	16.2	5.7	6.8	8.5	22.6	34.0	56.6
Hebei	16.0	5.6	6.7	8.4	22.3	33.5	55.8
Shanxi	16.1	5.6	6.7	8.4	22.4	33.7	56.1
Neimenggu	16.3	5.7	6.8	8.6	22.8	34.2	57.0
Liaoning	16.3	5.7	6.8	8.5	22.8	34.2	56.9
Jilin	16.2	5.7	6.8	8.5	22.6	33.9	56.6
Heilongjiang	16.0	5.6	6.7	8.4	22.3	33.5	55.8
Shanghai	15.8	5.5	6.6	8.3	22.0	33.1	55.1
Jiangsu	15.4	5.4	6.5	8.1	21.5	32.3	53.8
Zhejiang	15.6	5.5	6.6	8.2	21.8	32.8	54.6
Anhui	15.4	5.4	6.5	8.1	21.6	32.3	53.9
Fujian	15.5	5.4	6.5	8.1	21.7	32.6	54.3
Jiangxi	15.2	5.3	6.4	8.0	21.3	31.9	53.1

Continued

Province/ Municipality/ Autonomous region	Long-term inhalation rates (m³/d)	Short-term inhalation rates (L/min)					
		Sleeping	Sedentary behavior	Light intensity activity	Moderate intensity activity	High intensity activity	Very high intensity activity
Shandong	16.1	5.6	6.8	8.4	22.5	33.8	56.3
Henan	15.6	5.5	6.6	8.2	21.9	32.8	54.7
Hubei	15.8	5.5	6.6	8.3	22.1	33.1	55.2
Hunan	15.8	5.5	6.6	8.3	22.1	33.2	55.3
Guangdong	15.0	5.2	6.3	7.9	21.0	31.5	52.5
Guangxi	15.1	5.3	6.3	7.9	21.0	31.6	52.6
Hainan	15.3	5.3	6.4	8.0	21.4	32.1	53.4
Chongqing	14.8	5.2	6.2	7.8	20.7	31.1	51.8
Sichuan	15.4	5.4	6.4	8.1	21.5	32.2	53.7
Guizhou	15.6	5.5	6.6	8.2	21.8	32.8	54.6
Yunnan	15.5	5.4	6.5	8.1	21.6	32.4	54.1
Xizang	15.3	5.3	6.4	8.0	21.3	32.0	53.3
Shaanxi	15.1	5.3	6.4	7.9	21.2	31.8	52.9
Gansu	15.6	5.4	6.5	8.2	21.8	32.7	54.5
Qinghai	16.0	5.6	6.7	8.4	22.3	33.5	55.8
Ningxia	16.0	5.6	6.7	8.4	22.4	33.6	56.0
Xinjiang	16.3	5.7	6.8	8.5	22.7	34.1	56.8

Source: 2013 CEERHAPS-A.
* Median values.

(Beibei Wang, Zongshuang Wang, Yongjie Wei, Feifei Wang,

Xiaoli Duan)

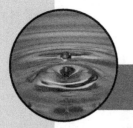

3 WATER INGESTION RATES

3.1 Introduction

Water ingestion rates include direct water ingestion rates (drinking of boiled water, raw water, barrel/bottled water, and water ingested from brewing coffee, tea, or drinking diluted powders), indirect water ingestion rates (the water ingested from porridge and soup), and total water ingestion rates (the sum of the direct and indirect water ingestion rates). They are expressed in units of water volume (ml) per day per capita.

Daily consumption of water may vary with different geographical or meteorological conditions (e.g., season, climate, and region) as well as lifestyles and food habits. The methods used to obtain water ingestion rates included measuring consumption and a questionnaire survey by 24-h dietary recall.

3.2 Recommended values

The recommended water ingestion rate values are based on the 2013 *Environmental exposure related activity patterns research for the Chinese population (Adults)* (CEERHAPS-A). The median value of daily consumption of water is a recommended value, which only represents water from local sources. Tables 3-1 to 3-3 provide the mean and 5th-95th percentile data for the daily total water ingestion rates, direct and indirect water ingestion rates by gender, age, urban/rural location , and region. Recommendations for different seasons are also provided. Table 3-4 provides the recommendations for water ingestion rate values by province.

Table 3-1 Recommended values for total water ingestion rates

Category		Daily total water ingestion * (ml/d)					
		Mean	P5	P25	P50	P75	P95
Total		2,300	638	1,203	1,850	2,785	5,200
Gender	Male	2,475	700	1,325	2,000	2,938	5,450
	Female	2,124	600	1,125	1,713	2,550	4,800
Age	18-44	2,315	650	1,238	1,875	2,800	5,250
	45-59	2,348	650	1,225	1,900	2,800	5,350
	60-79	2,220	600	1,138	1,800	2,700	4,850
	80+	1,898	450	938	1,525	2,300	4,500
Urban/Rural	Urban	2,355	675	1,250	1,900	2,800	5,325
	Rural	2,258	625	1,200	1,825	2,763	5,100
Region	North	2,856	900	1,650	2,338	3,356	5,975
	East	2,480	700	1,338	2,025	3,000	5,425
	South	1,995	663	1,174	1,650	2,325	4,300
	Northwest	2,595	700	1,400	2,100	3,075	6,025
	Northeast	1,551	500	875	1,275	1,950	3,125
	Southwest	1,973	451	938	1,500	2,300	5,000
Season	Spring/Autumn	2,159	550	1,100	1,710	2,650	4,925
	Summer	2,893	750	1,575	2,400	3,400	6,400
	Winter	1,990	500	950	1,600	2,400	4,700

Source: 2013 CEERHAPS-A.
* Milk, beverages, wine, and water in purchased food materials were not included.

Table 3-2 Recommended values for direct water ingestion rates

Category		Daily direct water ingestion (ml/d)					
		Mean	P5	P25	P50	P75	P95
Total		1,505	281	625	1,125	1,800	3,750
Gender	Male	1,638	313	750	1,250	2,000	4,000
	Female	1,372	250	625	1,000	1,625	3,375
Age	18-44	1,547	313	638	1,125	1,875	3,875
	45-59	1,529	275	625	1,125	1,875	3,750
	60-79	1,375	238	625	1,000	1,625	3,250
	80+	1,212	188	500	875	1,500	3,125
Urban/Rural	Urban	1,670	313	750	1,250	2,000	4,250
	Rural	1,378	250	625	1,100	1,688	3,250
Region	North	1,667	250	625	1,250	2,000	4,250
	East	1,632	313	688	1,225	2,000	4,000
	South	1,404	375	719	1,125	1,625	3,000
	Northwest	1,562	250	625	1,125	2,000	4,500
	Northeast	1,160	250	600	1,000	1,500	2,500
	Southwest	1,380	219	625	1,000	1,625	4,000
Season	Spring/Autumn	1,388	250	500	1,000	1,600	3,500
	Summer	2,015	400	1,000	1,500	2,500	5,000
	Winter	1,231	250	500	1,000	1,500	3,000

Source: 2013 CEERHAPS-A.

Table 3-3 Recommended values for indirect water ingestion rates

Category		Daily indirect water ingestion (ml/d)					
		Mean	P5	P25	P50	P75	P95
Total		799	100	250	480	960	2,400
Gender	Male	842	100	300	500	1,000	2,400
	Female	756	100	250	450	900	2,400
Age	18-44	772	100	250	450	900	2,400
	45-59	823	100	300	500	1,000	2,400
	60-79	850	100	300	500	1,050	2,400
	80+	692	100	250	450	800	1,950
Urban/Rural	Urban	687	100	250	400	800	2,000
	Rural	886	100	300	600	1,200	2,500
Region	North	1,196	200	400	800	1,600	3,200
	East	850	100	300	500	1,200	2,400
	South	593	88	200	400	750	1,800
	Northwest	1,034	200	400	800	1,350	3,200
	Northeast	391	0	200	350	450	900
	Southwest	603	50	250	400	800	1,600

Continued

Category		Daily indirect water ingestion (ml/d)					
		Mean	P5	P25	P50	P75	P95
Season	Spring/Autumn	775	50	200	400	800	2,400
	Summer	883	50	400	600	1,200	2,400
	Winter	763	40	200	400	800	2,400

Source: 2013 CEERHAPS-A.

Table 3-4 Recommended values for water ingestion rates by province [*]

Province/Municipality/Autonomous region	Daily total water ingestion (ml/d)				Daily direct water ingestion (ml/d)	Daily indirect water ingestion (ml/d)
	Annual average	Spring/Autumn	Summer	Winter		
Total	1,850	1,710	2,400	1,600	1,125	480
Beijing	2,325	2,200	2,700	2,200	1,875	400
Tianjin	2,125	1,900	2,500	1,900	1,563	400
Hebei	2,400	2,240	2,900	2,100	1,250	800
Shanxi	2,338	2,200	2,900	2,100	1,125	1,300
Neimenggu	1,925	1,800	2,300	1,800	1,500	400
Liaoning	1,325	1,300	1,500	1,300	1,000	400
Jilin	1,075	1,000	1,400	900	750	250
Heilongjiang	1,388	1,200	1,800	1,200	1,000	350
Shanghai	2,025	1,900	2,400	1,700	1,625	400
Jiangsu	1,900	1,800	2,400	1,700	1,125	600
Zhejiang	1,588	1,400	2,200	1,300	1,063	400
Anhui	2,650	2,370	3,280	2,300	1,125	1,000
Fujian	1,963	1,800	2,500	1,700	938	800
Jiangxi	1,431	1,350	2,000	1,150	1,125	300
Shandong	2,275	2,150	2,800	1,900	1,625	500
Henan	2,538	2,340	2,900	2,300	1,125	1,200
Hubei	1,625	1,400	2,300	1,200	1,125	400
Hunan	1,398	1,200	2,000	1,150	1,125	200
Guangdong	1,700	1,600	2,060	1,400	1,188	400
Guangxi	2,450	2,200	3,200	1,900	1,125	1,200
Hainan	1,400	1,400	1,700	1,150	625	600
Chongqing	1,050	900	1,400	900	625	400
Sichuan	1,625	1,400	2,400	1,300	1,125	500
Guizhou	1,275	1,200	1,800	950	938	250
Yunnan	1,650	1,550	1,950	1,400	1,125	400
Xizang	3,525	3,400	3,600	3,600	2,250	800
Shaanxi	1,900	1,800	2,400	1,700	1,125	500
Gansu	2,400	2,200	3,100	1,900	1,250	1,000

Continued

Province/Municipality/ Autonomous region	Daily total water ingestion (ml/d)				Daily direct water inges- tion (ml/d)	Daily indi- rect water ingestion (ml/d)
	Annual average	Spring/ Autumn	Summer	Winter		
Qinghai	1,900	1,850	1,900	1,800	1,200	400
Ningxia	1,950	1,800	2,400	1,550	1,375	400
Xinjiang	2,250	2,200	2,400	2,000	800	900

Source: 2013 CEERHAPS-A.
* Median values.

(Xiuge Zhao, Nan Huang, Chanjuan Zheng, Xianliang Wang,

Beibei Wang, Xiaoli Duan)

4 FOOD INTAKE

4.1 Introduction

The intake of food includes the consumption of different types of food. Food is classified into grains, vegetables, fruits, meat, fish and shrimp, dairy products, and eggs. The intake of food may vary with age, gender, economic condition, and lifestyle. The proportion of home-produced food (rice, wheat, vegetables, or fruits) consumed was calculated using the population of home-produced food consumed divided by the total population growing their own food.

4.2 Recommended values

The recommended values of food intake are based on the results of the China National Nutrition and Health Survey, which is organized periodically by the Ministry of Health. Four nation-wide nutrition surveys were conducted in 1959, 1982, 1992 and 2002. The 2002 data (Wang, 2005) were used in the recommendation for food intake. It is a nationally representative sample of 71,971 households (24,034 from urban areas and 47,937 from rural areas) or 270,000 samples from 31 provinces (Municipality or autonomous regions). Table 4-1 provides the mean values of food intake in units of grams per day (g/d) as the recommended values.

Table 4-1 Recommended values for food intake

Category		Food intake* (g/d)		
		Combined	Urban	Rural
Total food intake		1,056.6	1,117.7	1,033.0
Grain products	Rice products	238.3	217.8	246.2
	Wheat products	140.2	131.9	143.5
	Other grain products	23.6	16.3	26.4
Vegetables	Potato products	49.1	31.9	55.7
	Dark color vegetables	90.8	88.1	91.8
	Light color vegetables	185.4	163.8	193.8
Fruits		45.0	69.4	35.6
Meat products	Pork products	50.8	60.3	47.2
	Livestock meat products	9.2	15.5	6.8
	Poultry products	13.9	22.6	10.6
Fish and shellfish		29.6	44.9	23.7
Dairy products		26.5	65.8	11.4
Egg products		23.7	33.2	20.0

Source: 2002 China national nutrition and health survey.
* Results presented as daily intake for an adult male (age 18 years) participating in light physical activity.

Data from the 2013 *Environmental exposure related activity patterns research for the Chinese population (Adults)* (CEERHAPS-A) were analyzed to determine the recommended proportions of the population consuming home-produced foods, which were classified into grains (rice and wheat), vegetables, and fruits. Table 4-2 provides the proportion of the population consuming home-produced grain (rice and wheat), vegetables, and fruits, and Table 4-3 provides the proportion of the population consuming home-produced food by province.

Table 4-2 Proportion of population consuming home-produced food

Category		Proportion (%)		
		Combined	Urban	Rural
Total		97.8	97.1	98.0
Grain products	Rice	39.4	37.5	39.9
	Wheat	31.4	30.6	31.6
Vegetables		88.0	81.5	89.6
Fruits		25.0	20.4	26.2

Source: 2013 CEERHAPS-A.
* Proportion = population consuming specific home-produced food items divided by the total population growing food.

Table 4-3 Proportion of population consuming home-produced food by province

Province/Municipality/Autonomous region	Proportion of population consuming home-produced food (%)		
	Combined	Urban	Rural
Total	97.8	97.1	98.0
Beijing	82.2	73.3	86.1
Tianjin	99.1	98.2	100.0
Hebei	97.8	98.6	97.5
Shanxi	99.0	98.4	99.1
Neimenggu	96.9	94.7	97.1
Liaoning	96.8	89.0	98.3
Jilin	98.7	98.3	98.8
Heilongjiang	97.2	93.5	97.7
Shanghai	100.0	100.0	—
Jiangsu	98.5	98.1	98.7
Zhejiang	95.8	94.0	96.4
Anhui	94.9	97.2	94.6
Fujian	96.3	93.7	97.2
Jiangxi	98.8	98.5	99.0
Shandong	97.5	98.4	97.2
Henan	98.8	96.7	99.1
Hubei	98.6	97.8	98.9
Hunan	98.7	97.0	98.8
Guangdong	99.9	99.5	100.0
Guangxi	98.1	94.4	98.7
Hainan	98.1	90.7	99.1

Continued

Province/Municipality/ Autonomous region	Proportion of population consuming home-produced food * (%)		
	Combined	Urban	Rural
Chongqing	99.8	99.2	100.0
Sichuan	98.8	97.3	99.5
Guizhou	98.8	97.4	99.0
Yunnan	96.1	97.9	95.6
Xizang	99.7	100.0	99.6
Shaanxi	98.5	98.2	98.5
Gansu	95.9	99.3	95.1
Qinghai	98.3	94.8	99.0
Ningxia	98.6	99.0	97.6
Xinjiang	99.7	97.4	100.0

Source: 2013 CEERHAPS-A.
* Proportion = population consuming specific home-produced food items divided by the total population growing food.

It should be noted that the Ministry of Health periodically performs nutrition and health condition monitoring across the country. When conducting health risk assessments for oral exposure, the latest results of the nutrition survey should be considered.

(Yiting Chen, Suzhen Cao, Jing Nie, Xiuge Zhao, Beibei Wang,

Xiaoli Duan)

5 TIME-ACTIVITY FACTORS RELATED TO AIR EXPOSURE

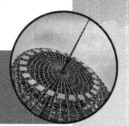

5.1 Introduction

Time-activity factors related to air exposure include the time spent indoors, time spent outdoors, and the time in transit. The categories of transit used in daily life were classified as walking, bicycle, electric bicycle, motorcycle, car, bus, and train.

Data for time-activity factors were usually obtained through questionnaires and by 24-h activity recall diaries. Other methods included the use of videotaping and GPS technology to provide location information.

5.2 Recommended values

The recommended values of time-activity factors related to air exposure for adults are based on the 2013 *Environmental Exposure Related Activity Patterns Research for the Chinese Population (Adults)* (CEERHAPS-A). Table 5-1 provides the mean and

5th-95th percentile data of time spent indoors by gender, age, urban/rural location, and region. Table 5-2 provides the mean and 5th-95th percentile data of time spent outdoors by gender, age, urban/rural location, and region. Tables 5-3 to 5-10 provide the mean and 5th-95th percentile data for time spent in transit and the time spent on different modes of transport by gender, age, urban/rural location, and region. Table 5-11 provides the recommended values of time spent indoors, time spent outdoors and the time spent in transit by province.

Table 5-1 Recommended values of time spent indoors

Category		Time indoors (min/d)					
		Mean	P5	P25	P50	P75	P95
Total		1,167	876	1,065	1,200	1,290	1,373
Gender	Male	1,152	855	1,043	1,185	1,283	1,370
	Female	1,183	900	1,095	1,215	1,300	1,377
Age	18-44	1,167	875	1,065	1,201	1,292	1,372
	45-59	1,157	866	1,051	1,185	1,285	1,370
	60-79	1,178	900	1,086	1,203	1,295	1,380
	80+	1,228	945	1,164	1,260	1,331	1,398
Urban/Rural	Urban	1,198	900	1,120	1,239	1,313	1,380
	Rural	1,142	865	1,035	1,165	1,269	1,364
Region	North	1,150	870	1,045	1,185	1,275	1,353
	East	1,197	900	1,110	1,239	1,315	1,383
	South	1,152	864	1,060	1,176	1,273	1,364
	Northwest	1,133	855	1,036	1,155	1,251	1,347
	Northeast	1,196	910	1,103	1,230	1,310	1,380
	Southwest	1,148	846	1,030	1,174	1,290	1,379

Source: 2013 CEERHAPS-A.

Table 5-2 Recommended values of time spent outdoors

Category		Time outdoors (min/d)					
		Mean	P5	P25	P50	P75	P95
Total		253	51	129	221	354	545
Gender	Male	267	53	135	236	377	561
	Female	239	50	121	209	330	525

Continued

Category		Time outdoors (min/d)					
		Mean	P5	P25	P50	P75	P95
Age	18-44	253	53	128	219	356	548
	45-59	264	56	136	235	369	559
	60-79	241	48	124	210	330	530
	80+	180	19	80	150	240	463
Urban/Rural	Urban	219	42	106	180	298	523
	Rural	279	60	150	255	385	555
Region	North	269	64	147	231	371	550
	East	225	40	109	183	311	520
	South	268	60	148	241	360	554
	Northwest	290	70	169	270	390	570
	Northeast	213	39	105	180	300	490
	Southwest	274	53	138	249	390	566
Season	Spring/Autumn	259	50	120	223	361	570
	Summer	295	55	150	260	420	630
	Winter	199	30	87	152	274	510

Source: 2013 CEERHAPS-A.

Table 5-3 Recommended values of time spent in transit

Category		Total (min/d)					
		Mean	P5	P25	P50	P75	P95
Total		63	15	30	45	70	180
Gender	Male	68	15	30	50	80	180
	Female	58	10	30	40	60	150
Age	18-44	64	15	30	50	75	180
	45-59	64	15	30	50	75	180
	60-79	59	10	30	40	65	160
	80+	54	10	20	30	60	160
Urban/Rural	Urban	66	15	30	50	80	180
	Rural	61	15	30	40	65	180
Region	North	61	20	30	50	70	160
	East	61	10	30	40	65	180
	South	64	15	30	50	80	180
	Northwest	84	20	30	60	120	200
	Northeast	57	10	30	40	60	150
	Southwest	61	10	30	40	60	180

Source: 2013 CEERHAPS-A.

Table 5-4 Recommended values of time spent walking

Category		Time spent walking (min/d)					
		Mean	P5	P25	P50	P75	P95
Total		52	10	30	30	60	120
Gender	Male	53	10	30	30	60	120
	Female	51	10	30	30	60	120
Age	18-44	50	10	20	30	60	120
	45-59	54	10	30	40	60	120
	60-79	54	10	30	40	60	120
	80+	49	10	20	30	60	120
Urban/Rural	Urban	52	10	30	30	60	120
	Rural	52	10	30	30	60	120
Region	North	48	10	30	30	60	120
	East	50	10	20	30	60	120
	South	50	10	30	30	60	120
	Northwest	70	20	30	60	120	180
	Northeast	43	10	20	30	60	120
	Southwest	54	10	20	30	60	180

Source: 2013 CEERHAPS-A.

Table 5-5 Recommended values of time spent traveling by bicycle

Category		Time spent traveling by bicycle (min/d)					
		Mean	P5	P25	P50	P75	P95
Total		41	10	20	30	60	120
Gender	Male	45	10	20	30	60	120
	Female	38	10	20	30	50	90
Age	18-44	40	10	20	30	60	120
	45-59	42	10	20	30	60	120
	60-79	40	10	20	30	50	120
	80+	38	10	30	30	40	80
Urban/Rural	Urban	43	10	20	30	60	120
	Rural	38	10	20	30	50	100
Region	North	43	10	20	30	60	120
	East	37	10	20	30	45	100
	South	45	10	20	30	60	120
	Northwest	43	10	20	30	60	120
	Northeast	39	10	20	30	40	120
	Southwest	33	0	15	30	40	80

Source: 2013 CEERHAPS-A.

Table 5-6 Recommended values of time spent traveling by electric bicycle

Category		Time spent traveling by electric bicycle (min/d)					
		Mean	P5	P25	P50	P75	P95
Total		43	10	20	30	60	120
Gender	Male	45	10	20	30	60	120
	Female	41	10	20	30	60	100
Age	18-44	43	10	20	30	60	120
	45-59	44	10	20	30	60	120
	60-79	43	0	20	30	60	120
	80+	80	0	20	30	80	480
Urban/Rural	Urban	43	10	20	30	60	120
	Rural	43	10	20	30	60	120
Region	North	45	10	20	35	60	120
	East	40	10	20	30	60	120
	South	48	15	30	40	60	120
	Northwest	46	10	20	30	60	120
	Northeast	54	10	30	30	60	120
	Southwest	35	0	20	30	45	60

Source: 2013 CEERHAPS-A.

Table 5-7 Recommended values of time spent traveling by motorcycle

Category		Time spent traveling by motorcycle (min/d)					
		Mean	P5	P25	P50	P75	P95
Total		46	10	20	30	60	120
Gender	Male	49	10	25	30	60	120
	Female	37	10	20	30	40	100
Age	18-44	46	10	20	30	60	120
	45-59	46	10	20	30	60	120
	60-79	36	0	20	30	40	110
	80+	43	0	20	30	60	120
Urban/Rural	Urban	49	10	20	30	60	120
	Rural	44	10	20	30	60	120
Region	North	43	10	20	30	60	120
	East	47	10	20	30	60	120
	South	46	10	30	30	60	120
	Northwest	53	10	28	30	60	120
	Northeast	44	10	20	30	60	120
	Southwest	44	10	20	30	60	120

Source: 2013 CEERHAPS-A.

Table 5-8 Recommended values of time spent traveling by car

Category		Time spent traveling by car (min/d)					
		Mean	P5	P25	P50	P75	P95
Total		71	10	30	40	60	240
Gender	Male	81	10	30	60	100	240
	Female	47	10	20	30	60	120
Age	18-44	73	10	30	40	70	240
	45-59	68	10	30	40	80	180
	60-79	52	10	20	30	60	180
	80+	35	10	30	40	40	60
Urban/Rural	Urban	69	10	30	40	60	200
	Rural	74	10	30	45	80	240
Region	North	72	10	30	40	60	200
	East	71	10	30	50	80	240
	South	68	10	30	40	60	240
	Northwest	78	10	20	40	80	330
	Northeast	63	10	20	40	80	180
	Southwest	73	10	20	40	80	240

Source: 2013 CEERHAPS-A.

Table 5-9 Recommended values of time spent traveling by bus

Category		Time spent traveling by bus (min/d)					
		Mean	P5	P25	P50	P75	P95
Total		43	10	20	30	60	120
Gender	Male	45	10	20	30	60	120
	Female	42	10	20	30	60	120
Age	18-44	45	10	20	30	60	120
	45-59	43	10	20	30	60	120
	60-79	37	10	20	30	45	90
	80+	44	10	20	30	60	100
Urban/Rural	Urban	45	10	20	30	60	120
	Rural	38	10	20	30	40	120
Region	North	47	10	20	30	60	120
	East	42	10	20	30	60	120
	South	47	10	20	30	60	120
	Northwest	43	10	20	30	60	120
	Northeast	46	10	30	30	60	120
	Southwest	34	10	15	30	40	100

Source: 2013 CEERHAPS-A.

Table 5-10 Recommended values of time spent traveling by train

Category		Time spent traveling by train (min/d)					
		Mean	P5	P25	P50	P75	P95
Total		40	5	20	30	50	120
Gender	Male	43	5	20	30	50	180
	Female	38	8	20	30	40	120
Age	18-44	47	5	20	30	50	180
	45-59	31	5	15	30	30	60
	60-79	38	5	20	30	30	120
	80+	32	10	20	30	30	60
Urban/Rural	Urban	39	5	20	30	50	120
	Rural	66	20	20	30	120	180
Region	North	41	7	20	30	40	120
	East	53	5	20	30	60	180
	South	33	8	20	30	40	60
	Northwest	59	1	1	2	120	120
	Northeast	93	1	20	80	180	180
	Southwest	18	10	10	15	30	30

Source: 2013 CEERHAPS-A.

Table 5-11 Recommended values of time-activity factors related to air exposure by province [*]

Category	Time spent indoors (min/d)	Time spent outdoors (min/d)					Time spent in transit (min/d)						
		Annual average	Spring/Autumn	Summer	Winter	Total	Walking	Bicycle	Electric bicycle	Motorcycle	Car	Bus	Train
Total	1,200	221	223	260	152	45	30	30	30	30	40	30	30
Beijing	1,220	195	197	240	143	60	50	30	40	50	60	40	30
Tianjin	1,204	188	180	244	120	60	50	30	40	30	60	60	30
Hebei	1,233	182	180	245	120	40	30	40	40	30	40	40	240
Shanxi	1,190	225	230	300	120	40	35	30	30	30	60	40	20
Neimenggu	1,174	250	226	350	126	60	30	30	30	30	60	40	0
Liaoning	1,230	185	201	240	90	40	30	30	50	30	50	30	180
Jilin	1,273	146	133	220	60	45	40	30	20	30	40	40	40
Heilongjiang	1,202	210	210	300	57	40	30	30	20	30	40	30	0
Shanghai	1,264	155	154	169	122	40	30	30	30	30	30	30	30
Jiangsu	1,260	170	180	180	137	40	30	30	30	30	45	30	30
Zhejiang	1,230	192	197	190	166	40	30	30	30	30	40	30	20
Anhui	1,210	206	206	231	179	60	40	30	40	40	60	30	180
Fujian	1,241	186	180	209	173	50	30	30	30	30	60	20	30
Jiangxi	1,196	227	234	249	189	50	30	30	30	30	60	30	0
Shandong	1,247	169	164	219	110	40	30	30	30	30	40	30	5
Henan	1,125	300	300	360	230	50	30	30	30	40	40	30	10
Hubei	1,166	265	270	300	219	60	40	50	50	60	40	30	30
Hunan	1,208	210	220	240	160	35	30	30	30	30	40	30	30

Continued

Category	Time spent indoors (min/d)	Time spent outdoors (min/d)				Time spent in transit (min/d)							
		Annual average	Spring/Autumn	Summer	Winter	Total	Walking	Bicycle	Electric bicycle	Motorcycle	Car	Bus	Train
Guangdong	1,200	210	210	240	180	40	30	30	30	30	30	60	30
Guangxi	1,140	290	280	349	231	50	40	30	30	38	60	30	40
Hainan	1,097	334	330	360	300	40	30	30	30	30	60	30	0
Chongqing	1,235	207	200	240	140	30	20	10	60	20	40	18	0
Sichuan	1,170	249	253	270	214	40	40	30	30	30	60	30	15
Guizhou	1,195	229	239	271	150	60	40	35	30	30	40	30	0
Yunnan	1,095	320	326	360	270	40	30	25	30	30	30	20	0
Xizang	1,067	339	360	350	283	90	60	30	30	50	30	40	0
Shaanxi	1,157	270	283	326	190	60	60	40	30	30	30	30	2
Gansu	1,095	330	330	429	197	60	60	30	30	30	30	30	120
Qinghai	1,230	169	154	240	120	50	30	30	15	30	30	30	1
Ningxia	1,247	165	171	201	120	40	40	30	30	30	60	30	0
Xinjiang	1,145	285	286	360	187	90	60	30	40	60	60	30	0

Source: 2013 CEERHAPS-A.
* Median values.

(Beibei Wang, Suzhen Cao, Jin Ma, Nan Huang, Jing Nie, Zongshuang Wang, Xiaoli Duan)

6 TIME-ACTIVITY FACTORS RELATED TO WATER EXPOSURE

6.1 Introduction

Time-activity factors related to water exposure refer to the time that a body is in direct contact with water, including time spent bathing/showering and swimming.

6.2 Recommended values

Data recommended in this chapter are based on the 2013 *Environmental exposure related activity patterns research for the Chinese population (Adults)* (CEERHAPS-A). Activity factors in this survey were obtained through questionnaires by 24-h recall activity diaries. The mean and 5th-95th percentile data by gender, age, urban/rural location, and region for time spent showering/bathing are summarized in Table 6-1. Table 6-2 provides the percentages of the population that swims and the 5th-95th percentile data of the time spent swimming for participants by gender, age, urban/rural location, and region. Table 6-3 provides the recommended values of the time spent showering/bathing and the time spent swimming for swimmers by province. The time spent showering/bathing is reported in units of minutes per day, while the time spent swimming is reported in units of minutes per month.

Table 6-1 Recommended values of time spent showering/bathing

Category		Time spent showering/bathing * (min/d)					
		Mean	P5	P25	P50	P75	P95
Total		8	1	4	7	10	18
Gender	Male	8	1	4	7	10	18
	Female	8	1	4	7	11	19
Age	18-44	9	2	5	8	11	19
	45-59	8	1	4	7	10	18
	60-79	7	1	3	6	9	15
	80+	6	0	2	5	8	16
Urban/Rural	Urban	9	2	5	8	11	20
	Rural	7	1	4	6	10	16
Region	North	7	1	3	5	9	16
	East	10	3	6	9	12	20
	South	10	4	7	10	13	20
	Northwest	5	0	2	4	6	12
	Northeast	7	1	3	6	9	16
	Southwest	6	1	4	5	8	14
Season	Spring/Autumn	7	1	3	5	10	20
	Summer	13	2	7	10	15	30
	Winter	6	0	2	4	8	15

Source: 2013 CEERHAPS-A.

* Time spent showering/bathing is the time spent on either showering or bathing.

Table 6-2 Proportion of the population that swims and the recommended values of the time spent swimming

Category		Proportion of the population that swims (%)	Time spent swimming (min/month)					
			Mean	P5	P25	P50	P75	P95
Total		3.3	155	8	30	75	180	593
Gender	Male	5.3	154	8	30	75	173	550
	Female	1.5	159	0	30	80	185	630

Continued

Category		Proportion of the popula-tion that swims (%)	Time spent swimming* (min/month)					
			Mean	P5	P25	P50	P75	P95
Age	18-44	5.5	148	8	30	75	170	510
	45-59	2.3	181	0	30	75	210	750
	60-79	1.0	169	0	30	75	150	720
	80+	0.3	117	75	105	105	120	195
Urban/Rural	Urban	4.0	180	8	30	80	200	720
	Rural	2.6	123	5	23	60	150	450
Region	North	1.5	143	8	30	60	165	600
	East	3.0	196	10	30	75	225	875
	South	5.0	158	8	30	90	180	600
	Northwest	2.3	75	8	15	40	90	240
	Northeast	1.7	132	0	30	65	150	480
	Southwest	6.1	134	5	30	75	175	450
Season	Spring/Autumn		64	0	0	0	20	300
	Summer	—	459	15	100	240	600	1,800
	Winter		30	0	0	0	0	150

Source: 2013 CEERHAPS-A.
* Swimmers only.

Table 6-3 Recommended values of time spent showering/bathing and time spent swimming by province

Province/Municipality/Autonomous region	Time showering/bathing*[1] (min/d)				Proportion of swimming population (%)	Time swimming*[2] (min/month)			
	Annual average	Spring/Autumn	Summer	Winter		Annual average	Spring/Autumn	Summer	Winter
Total	7	5	10	4	3.3	155	64	459	30
Beijing	9	8	15	5	4.2	275	246	510	171
Tianjin	9	7	15	3	0.8	197	98	488	35
Hebei	6	4	10	3	1.3	158	59	465	51
Shanxi	3	2	4	2	1.2	149	143	295	11
Neimenggu	3	2	4	2	0.7	126	103	222	31
Liaoning	8	5	13	4	2.2	100	70	208	50
Jilin	5	4	7	3	1.4	108	76	236	44
Heilongjiang	4	3	8	2	1.5	193	168	324	111
Shanghai	10	10	15	7	4.0	193	106	493	67
Jiangsu	10	8	15	7	2.7	211	58	698	32
Zhejiang	9	8	13	5	4.0	157	29	555	9
Anhui	9	5	15	5	0.9	110	17	403	0
Fujian	10	10	15	8	4.0	309	149	862	75
Jiangxi	8	7	10	5	3.6	113	28	399	8

Continued

Province/ Municipality/ Autonomous region	Time showering/bathing[*1] (min/d)				Proportion of swimming population (%)	Time swimming[**2] (min/month)			
	Annual average	Spring/ Autumn	Summer	Winter		Annual average	Spring/ Autumn	Summer	Winter
Shandong	7	4	13	4	2.9	172	50	559	30
Henan	6	3	11	3	1.9	81	36	230	20
Hubei	8	6	10	5	4.0	160	32	556	19
Hunan	8	8	10	5	6.5	111	25	391	3
Guangdong	10	10	10	10	3.8	188	105	506	49
Guangxi	10	10	10	10	6.0	266	192	599	12
Hainan	14	13	15	10	2.7	164	168	260	62
Chongqing	4	3	10	2	0.6	30	0	119	0
Sichuan	6	4	10	3	1.5	120	21	437	0
Guizhou	6	4	10	3	8.3	165	29	597	3
Yunnan	5	4	6	3	6.5	119	89	231	63
Xizang	0	0	0	0	18.2	39	44	53	2
Shaanxi	5	4	10	3	0.8	165	45	556	15
Gansu	2	2	3	1	0.5	93	28	271	45
Qinghai	4	3	5	3	0.9	159	101	352	83
Ningxia	6	5	8	4	4.7	159	122	307	81
Xinjiang	4	4	5	3	5.7	49	4	177	3

Source: 2013 CEERHAPS-A.
* Median values.
** Mean values.
1. Time showering/bathing is the time spent either showering or bathing.
2. Swimmers only.

(Xiuge Zhao, Yiting Chen, Beibei Wang, Suzhen Cao,

Delong Fan, Xiaoli Duan)

7 TIME-ACTIVITY FACTORS RELATED TO SOIL EXPOSURE

7.1 Introduction

Exposure to soil may occur during farming, working related activities, outdoor sports or recreational activities. Farming refers to activities undertaken during agricultural production, in which gardening work is not included. Working related processes refers to activities that entail contact with soil during work (e.g., construction). The time in contact with soil is expressed as a time-activity factor related to soil exposure, with units of minutes per day.

7.2 Recommended values

The recommended values of time-activity factors related to soil exposure for adults are based on the 2013 *Environmental exposure related activity patterns research for the Chinese population (Adults)* (CEERHAPS-A).

Table 7-1 provides the proportion of the population that
come into contact with soil, and the mean and 5th-95th percentile data of the contact time. Tables 7-2 to 7-4 provide the mean
and 5th-95th percentile data of time in contact with soil during
farming, working related processes, outdoor sports or recreational activities by gender, age, urban/rural location, and region
respectively. Table 7-5 provides the mean and 5th-95th percentile data for the time in contact with soil by province.

Table 7-1 Recommended values of time in contact with soil [*]

Category		Proportion of population with soil contact activities (%)	Time in contact with soil [**] (min/d)					
			Mean	P5	P25	P50	P75	P95
Total		47.1	204	20	75	180	300	480
Gender	Male	48.5	212	20	90	180	300	480
	Female	46.0	195	20	60	180	300	480
Age	18-44	44.4	205	20	90	180	300	480
	45-59	53.0	215	30	90	180	300	480
	60-79	44.7	183	20	60	130	240	480
	80+	18.2	112	10	30	60	180	360
Urban/Rural	Urban	21.6	168	10	60	120	240	480
	Rural	68.7	214	30	120	180	300	480
Region	North	47.4	217	30	120	180	300	480
	East	37.6	177	15	60	120	240	480
	South	41.5	199	30	90	180	300	480
	Northwest	57.0	230	30	120	190	300	500
	Northeast	50.4	229	30	60	180	360	600
	Southwest	58.5	198	20	60	180	300	480

Source: 2013 CEERHAPS-A.
* Includes soil and outdoor settled dust.
** Participants only, includes time in contact with soil while farming, working or doing outdoor sports.

Table 7-2 Recommended values of time in contact with soil during farming [*]

Category		Time in contact with soil during farming [**] (min/d)					
		Mean	P5	P25	P50	P75	P95
Total		201	25	90	180	300	480
Gender	Male	208	30	90	180	300	480
	Female	194	20	70	180	300	480

Continued

Category		Time in contact with soil during farming ** (min/d)					
		Mean	P5	P25	P50	P75	P95
Age	18-44	201	30	90	180	300	480
	45-59	212	30	120	180	300	480
	60-79	182	20	60	150	240	480
	80+	115	10	30	60	180	360
Urban/Rural	Urban	173	20	60	120	240	480
	Rural	208	30	120	180	300	480
Region	North	214	30	120	180	300	480
	East	175	15	60	120	240	480
	South	195	30	100	180	270	480
	Northwest	229	60	120	180	300	480
	Northeast	230	30	90	180	360	600
	Southwest	196	20	60	180	300	480

Source: 2013 CEERHAPS-A.
* Includes soil and outdoor settled dust.
** Participants only.

Table 7-3 Recommended values of time in contact with soil during working related processes *

Category		Time in contact with soil during working related processes ** (min/d)					
		Mean	P5	P25	P50	P75	P95
Total		140	10	60	90	180	480
Gender	Male	163	15	60	120	240	480
	Female	111	10	60	60	120	360
Age	18-44	155	10	60	120	180	480
	45-59	137	15	60	90	121	480
	60-79	112	10	40	60	120	420
	80+	70	20	20	60	90	120
Urban/Rural	Urban	127	15	60	60	120	480
	Rural	146	10	60	120	180	480
Region	North	185	15	60	120	300	480
	East	163	10	40	120	240	480
	South	111	30	60	60	120	360
	Northwest	123	30	60	90	150	360
	Northeast	117	30	60	120	120	360
	Southwest	134	0	30	60	180	480

Source: 2013 CEERHAPS-A.
* Includes soil and outdoor settled dust.
** Participants only.

Table 7-4 Recommended values of time in contact with soil during outdoor sports[*]

Category		Time in contact with soil during outdoor sports [**] (min/d)					
		Mean	P5	P25	P50	P75	P95
Total		64	2	30	60	60	180
Gender	Male	64	2	30	60	80	180
	Female	63	0	20	60	60	200
Age	18-44	60	0	20	60	60	180
	45-59	65	5	30	60	60	180
	60-79	70	5	30	60	90	240
	80+	55	0	10	30	60	180
Urban/Rural	Urban	50	5	20	30	60	120
	Rural	74	0	30	60	120	240
Region	North	76	10	30	60	110	240
	East	59	3	30	60	60	123
	South	52	5	20	30	60	180
	Northwest	59	2	30	60	60	120
	Northeast	54	3	15	30	60	180
	Southwest	63	0	10	30	60	240

Source: 2013 CEERHAPS-A.
* Includes soil and outdoor settled dust.
** Participants only.

Table 7-5 Recommended values of time in contact with soil by province[*]

Province/Municipality/ Autonomous region	Proportion of population in contact with soil (%)	Time in contact with soil [**] (min/d)			
		Combined	Farming	Working related processes	Outdoor sports
Total	47.1	204	201	140	64
Beijing	24.2	146	163	236	52
Tianjin	37.2	220	242	70	71
Hebei	49.2	198	194	178	80
Shanxi	54.5	239	228	214	59
Neimenggu	43.0	286	284	116	46
Liaoning	59.9	238	240	121	49
Jilin	39.2	160	157	119	59
Heilongjiang	50.1	266	270	100	50
Shanghai	5.4	68	78	62	26
Jiangsu	36.0	122	123	131	40
Zhejiang	38.4	148	142	155	29
Anhui	36.8	182	178	137	74
Fujian	32.8	223	220	256	14
Jiangxi	37.9	146	140	101	44
Shandong	47.7	205	203	211	69

Continued

Province/Municipality/ Autonomous region	Proportion of population in contact with soil (%)	Time in contact with soil ** (min/d)			
		Combined	Farming	Working related processes	Outdoor sports
Henan	51.1	209	202	265	85
Hubei	29.1	193	192	93	44
Hunan	52.3	161	156	128	62
Guangdong	35.4	220	205	105	55
Guangxi	44.2	219	218	112	28
Hainan	49.6	234	234	250	193
Chongqing	48.7	119	121	570	36
Sichuan	50.3	183	181	117	73
Guizhou	62.1	209	212	134	54
Yunnan	70.0	246	246	202	87
Xizang	56.1	236	211	51	5
Shaanxi	58.7	264	247	144	71
Gansu	79.2	271	271	90	50
Qinghai	55.5	227	220	129	8
Ningxia	26.3	162	170	179	53
Xinjiang	46.0	188	191	138	51

Source: 2013 CEERHAPS-A.
* Average time spent by participants.
** Includes soil and outdoor settled dust.

(Beibei Wang, Suzhen Cao, Xiuge Zhao, Ting Dong, Jing Nie,

Xiaoli Duan)

8 TIME-ACTIVITY FACTORS RELATED TO ELECTROMAGNETIC EXPOSURE

8.1 Introduction

Time-activity factors related to electromagnetic exposure should be considered when people use electronic devices. In this chapter, it refers largely to time spent using computers or talking on cell phones, and is expressed in units of minutes per day.

8.2 Recommended values

The recommended values of time-activity factors related to electromagnetic exposure for adults are based on the 2013 *Environmental exposure related activity patterns research for the Chinese population (Adults)* (CEERHAPS-A). Table 8-1 provides the proportion of the population using a computer and the 5th-95th percentile data of the time spent using a computer for users by gender, age, urban/rural location, and region. Table 8-2 provides the proportion of the population using a cell phone and the 5th-95th percentile data for time spent talking on cell phones for users by gender, age, urban/rural location, and region. Table 8-3 provides the proportion of users, and the recommended values of the time spent using a computer and talking on a cell phone by province.

Table 8-1 Recommended values of time spent using a computer

Category		Proportion of population using a computer (%)	Time spent using a computer * (min/d)					
			Mean	P5	P25	P50	P75	P95
Total		29.5	167	30	60	120	240	480
Gender	Male	32.1	162	20	60	120	240	480
	Female	27.3	173	30	60	120	240	480
Age	18-44	47.2	175	30	60	120	240	480
	45-59	22.8	144	20	60	120	180	380
	60-79	10.1	138	20	60	120	180	360
	80+	4.6	146	20	60	120	180	480
Urban/ Rural	Urban	43.2	188	30	60	120	240	480
	Rural	17.9	134	20	60	120	180	360
Region	North	27.7	151	30	60	120	180	480
	East	31.7	177	30	60	120	240	480
	South	34.0	163	20	60	120	210	480
	Northwest	26.5	161	30	60	120	210	480
	Northeast	35.8	169	30	60	120	240	480
	Southwest	20.7	174	20	60	120	240	480

Source: 2013 CEERHAPS-A.
* Users only.

Table 8-2 Recommended values of time spent talking on a cell phone

Category		Proportion of population using a cell phone (%)	Time spent talking on a cell phone * (min/d)					
			Mean	P5	P25	P50	P75	P95
Total		76.4	24	2	8	15	30	60
Gender	Male	79.9	26	3	10	15	30	70
	Female	73.6	22	2	5	10	30	60
Age	18-44	89.1	28	3	10	20	30	90
	45-59	78.5	20	2	5	10	20	60
	60-79	54.3	14	2	5	10	15	35
	80+	23.7	12	1	5	10	10	30

Continued

Category		Proportion of population using a cell phone (%)	Time spent talking on a cell phone* (min/d)					
			Mean	P5	P25	P50	P75	P95
Urban/ Rural	Urban	83.2	28	3	10	20	30	90
	Rural	70.6	21	2	5	10	20	60
Region	North	76.6	24	2	5	10	30	80
	East	76.6	25	3	10	15	30	60
	South	75.9	26	3	10	15	30	60
	Northwest	77.0	24	3	8	15	30	60
	Northeast	78.4	21	2	5	10	30	60
	Southwest	74.5	22	2	6	10	30	60

Source: 2013 CEERHAPS-A.
* Users only.

Table 8-3 Recommended values of time spent using a computer or cell phone by province

Province/ Municipality/ Autonomous region	Proportion of population using a computer (%)	Time spent using a computer* (min/d)	Proportion of population using a cell phone (%)	Time spent talking on a cell phone* (min/d)
Total	29.5	167	76.4	24
Beijing	45.9	206	74.2	33
Tianjin	40.4	168	69.8	38
Hebei	28.9	170	73.8	28
Shanxi	28.4	120	76.8	18
Neimenggu	20.0	121	89.8	19
Liaoning	22.6	171	69.9	24
Jilin	39.2	140	85.3	19
Heilongjiang	44.4	184	80.8	20
Shanghai	50.8	230	76.1	31
Jiangsu	28.9	197	71.9	23
Zhejiang	35.9	191	81.6	27
Anhui	25.9	153	69.2	24
Fujian	34.9	172	86.8	26
Jiangxi	36.1	163	84.2	27
Shandong	26.6	164	72.1	23
Henan	23.7	134	73.1	19
Hubei	56.9	171	77.7	29
Hunan	26.4	167	79.4	24
Guangdong	30.9	162	75.1	26
Guangxi	28.8	137	71.4	22
Hainan	16.0	139	72.7	26
Chongqing	23.3	184	74.0	20

Continued

Province/ Municipality/ Autonomous region	Proportion of population using a computer (%)	Time spent using a computer* (min/d)	Proportion of population using a cell phone (%)	Time spent talking on a cell phone* (min/d)
Sichuan	25.1	160	74.1	24
Guizhou	15.5	193	71.0	21
Yunnan	15.7	182	80.8	22
Xizang	26.9	125	68.7	35
Shaanxi	26.5	139	75.7	27
Gansu	28.7	141	80.6	18
Qinghai	27.3	177	83.6	33
Ningxia	46.8	184	83.9	27
Xinjiang	15.4	179	67.9	23

Source: 2013 CEERHAPS-A.
* Average time spent by users.

(Yan Qian, Beibei Wang, Yiting Chen, Xiuge Zhao,

Xiaoli Duan)

9 BODY WEIGHT

9.1 Introduction

Body weight is an important exposure factor to reflect the physical characteristics of a target population, and is used to normalize the average daily dose during exposure assessment. Body weight may be affected by genetic factors, economic conditions, and lifestyles.

9.2 Recommended values

The data presented here are based on the 2013 *Environmental exposure related activity patterns research for the Chinese population (Adults)* (CEERHAPS-A). Median values are used as the recommended values in kilograms. Table 9-1 provides the mean and 5th-95th percentile data of body weight by gender, age, urban/ rural location, and region. Table 9-2 provides the recommended values of body weight by province.

Table 9-1 Recommended values of body weight

Category		Body weight (kg)					
		Mean	P5	P25	P50	P75	P95
Total		61.9	45.1	53.6	60.6	69.0	82.7
Gender	Male	66.1	49.1	57.7	65.0	73.1	87.0
	Female	57.8	43.3	50.6	56.8	63.9	75.5
Age	18-44	61.9	45.5	53.0	60.1	68.8	83.8
	45-59	63.5	46.9	55.6	62.4	70.2	83.0
	60-79	60.3	43.1	52.4	59.4	67.5	80.0
	80+	55.5	38.8	47.6	54.3	62.5	75.1
Urban/Rural	Urban	63.4	46.3	54.8	62.0	70.6	84.8
	Rural	60.8	44.4	52.7	59.7	67.6	81.0
Region	North	65.5	48.8	57.5	64.3	72.3	86.4
	East	62.3	45.6	53.9	61.0	69.3	83.1
	South	58.6	43.0	50.9	57.3	65.0	78.0
	Northwest	62.6	46.2	54.6	61.5	69.4	82.9
	Northeast	65.6	48.5	57.1	64.0	72.3	87.6
	Southwest	58.3	43.5	51.0	57.1	64.6	76.9

Source: 2013 CEERHAPS-A.

Table 9-2 Recommended values of body weight by province *

Province/Municipality/ Autonomous region	Body weight (kg)	Province/Municipality/ Autonomous region	Body weight (kg)
Total	60.6	Henan	62.8
Beijing	66.9	Hubei	60.1
Tianjin	65.7	Hunan	57.3
Hebei	65.1	Guangdong	57.0
Shanxi	64.0	Guangxi	55.0
Neimenggu	64.8	Hainan	54.0
Liaoning	65.0	Chongqing	57.3
Jilin	63.5	Sichuan	58.2
Heilongjiang	63.3	Guizhou	56.0
Shanghai	62.2	Yunnan	55.9
Jiangsu	62.0	Xizang	55.1
Zhejiang	59.5	Shaanxi	59.0
Anhui	60.5	Gansu	61.8
Fujian	57.4	Qinghai	62.0
Jiangxi	55.9	Ningxia	62.7
Shandong	65.0	Xinjiang	62.4

Source: 2013 CEERHAPS-A.
* Median values.

(Suzhen Cao, Xiuge Zhao, Limin Wang, Beibei Wang,

Yiting Chen, Xiaoli Duan)

10 BODY SURFACE AREA

10.1 Introduction

Dermal exposure can occur during various activities in different environmental media (USEPA, 1992, 2011). The dose may be affected by dermal adherence to the skin, the film thickness of liquids on the skin, body surface area, and other factors. Here, recommended values are presented only in terms of body surface area because other factors are unavailable for China. Body surface area is provided as head, trunk and upper limbs (including arms and hands) and lower limbs (including legs and feet).

Surface area can be identified by measurement, which includes a coating method, triangulation, and surface integration, or it can be calculated by height and weight based on models. In this chapter, the value of surface area was calculated based on a formula (10-1) established by 3D measurement (HRPAC, 2008):

$$SA = 0.012\, H^{0.6} \times W^{0.45} \qquad (10\text{-}1)$$

In the formula, SA: Surface area, m^2; BW: body weight, kg; H: height, cm.

10.2 Recommended values

The recommended values are medians in units of m^2, and are based on the analysis of data from the 2013 *Environmental exposure related activity patterns research for the Chinese population (Adults)* (CEERHAPS-A). The recommended surface area of the whole body and different parts of the body is summarized in Tables 10-1 to 10-7, and is provided as a mean and 5th-95th percentile data by gender, age, urban/rural location, and region. Table 10-8 provides the recommended values of surface area by province.

Table 10-1 Recommended values of total body surface area

Category		Total surface area (m^2)					
		Mean	P5	P25	P50	P75	P95
Total		1.6	1.4	1.5	1.6	1.7	1.9
Gender	Male	1.7	1.4	1.6	1.7	1.8	2.0
	Female	1.5	1.3	1.4	1.5	1.6	1.8
Age	18-44	1.6	1.4	1.5	1.6	1.7	1.9
	45-59	1.6	1.4	1.5	1.6	1.7	1.9
	60-79	1.6	1.3	1.5	1.6	1.7	1.9
	80+	1.5	1.2	1.4	1.5	1.6	1.8
Urban/Rural	Urban	1.6	1.4	1.5	1.6	1.8	1.9
	Rural	1.6	1.3	1.5	1.6	1.7	1.9
Region	North	1.7	1.4	1.6	1.7	1.8	2.0
	East	1.6	1.4	1.5	1.6	1.7	1.9
	South	1.6	1.3	1.5	1.6	1.7	1.9
	Northwest	1.6	1.4	1.5	1.6	1.7	1.9
	Northeast	1.7	1.4	1.6	1.7	1.8	2.0
	Southwest	1.6	1.3	1.5	1.6	1.7	1.8

Source: 2013 CEERHAPS-A.

Table 10-2 Recommended values of the surface area of body parts—head

Category		Head (m²)					
		Mean	P5	P25	P50	P75	P95
Total		0.12	0.10	0.11	0.12	0.13	0.15
Gender	Male	0.13	0.11	0.12	0.13	0.14	0.15
	Female	0.12	0.10	0.11	0.12	0.12	0.14
Age	18-44	0.12	0.10	0.11	0.12	0.13	0.15
	45-59	0.12	0.10	0.12	0.12	0.13	0.15
	60-79	0.12	0.10	0.11	0.12	0.13	0.14
	80+	0.11	0.09	0.10	0.11	0.12	0.14
Urban/Rural	Urban	0.12	0.10	0.12	0.12	0.13	0.15
	Rural	0.12	0.10	0.11	0.12	0.13	0.14
Region	North	0.13	0.11	0.12	0.13	0.13	0.15
	East	0.12	0.10	0.11	0.12	0.13	0.15
	South	0.12	0.10	0.11	0.12	0.13	0.14
	Northwest	0.12	0.10	0.11	0.12	0.13	0.15
	Northeast	0.13	0.11	0.12	0.13	0.14	0.15
	Southwest	0.12	0.10	0.11	0.12	0.13	0.14

Source: 2013 CEERHAPS-A.

Table 10-3 Recommended values of the surface area of body parts—trunk

Category		Trunk (m²)					
		Mean	P5	P25	P50	P75	P95
Total		0.60	0.51	0.56	0.60	0.65	0.72
Gender	Male	0.63	0.54	0.59	0.63	0.67	0.74
	Female	0.57	0.49	0.54	0.57	0.61	0.66
Age	18-44	0.61	0.51	0.56	0.61	0.65	0.72
	45-59	0.61	0.51	0.57	0.61	0.65	0.72
	60-79	0.59	0.49	0.55	0.59	0.63	0.70
	80+	0.56	0.46	0.52	0.56	0.60	0.68
Urban/Rural	Urban	0.61	0.52	0.57	0.61	0.66	0.73
	Rural	0.60	0.50	0.55	0.60	0.64	0.71
Region	North	0.62	0.53	0.58	0.62	0.66	0.73
	East	0.61	0.51	0.56	0.61	0.65	0.72
	South	0.59	0.49	0.54	0.59	0.63	0.69
	Northwest	0.61	0.51	0.57	0.61	0.65	0.71
	Northeast	0.63	0.53	0.58	0.63	0.67	0.74
	Southwest	0.58	0.49	0.54	0.58	0.62	0.68

Source: 2013 CEERHAPS-A.

Table 10-4 Recommended values of the surface area of body parts—arms

Category		Arms (m²)					
		Mean	P5	P25	P50	P75	P95
Total		0.24	0.20	0.22	0.24	0.26	0.28
Gender	Male	0.25	0.21	0.23	0.25	0.27	0.29
	Female	0.23	0.19	0.21	0.23	0.24	0.26
Age	18-44	0.24	0.20	0.22	0.24	0.26	0.29
	45-59	0.24	0.20	0.23	0.24	0.26	0.28
	60-79	0.23	0.19	0.22	0.23	0.25	0.28
	80+	0.22	0.18	0.20	0.22	0.24	0.27
Urban/Rural	Urban	0.24	0.20	0.22	0.24	0.26	0.29
	Rural	0.24	0.20	0.22	0.24	0.25	0.28
Region	North	0.25	0.21	0.23	0.25	0.26	0.29
	East	0.24	0.20	0.22	0.24	0.26	0.28
	South	0.23	0.20	0.22	0.23	0.25	0.27
	Northwest	0.24	0.20	0.22	0.24	0.26	0.28
	Northeast	0.25	0.21	0.23	0.25	0.26	0.29
	Southwest	0.23	0.19	0.21	0.23	0.25	0.27

Source: 2013 CEERHAPS-A.

Table 10-5 Recommended values of the surface area of body parts—hands

Category		Hands (m²)					
		Mean	P5	P25	P50	P75	P95
Total		0.08	0.07	0.07	0.08	0.08	0.09
Gender	Male	0.08	0.07	0.08	0.08	0.09	0.09
	Female	0.07	0.06	0.07	0.07	0.08	0.09
Age	18-44	0.08	0.07	0.07	0.08	0.08	0.09
	45-59	0.08	0.07	0.07	0.08	0.08	0.09
	60-79	0.08	0.06	0.07	0.08	0.08	0.09
	80+	0.07	0.06	0.07	0.07	0.08	0.09
Urban/Rural	Urban	0.08	0.07	0.07	0.08	0.08	0.09
	Rural	0.08	0.06	0.07	0.08	0.08	0.09
Region	North	0.08	0.07	0.07	0.08	0.09	0.09
	East	0.08	0.07	0.07	0.08	0.08	0.09
	South	0.08	0.06	0.07	0.08	0.08	0.09
	Northwest	0.08	0.07	0.07	0.08	0.08	0.09
	Northeast	0.08	0.07	0.07	0.08	0.09	0.10
	Southwest	0.07	0.06	0.07	0.07	0.08	0.09

Source: 2013 CEERHAPS-A.

Table 10-6 Recommended values of the surface area of body parts—legs

Category		Legs (m²)					
		Mean	P5	P25	P50	P75	P95
Total		0.47	0.39	0.43	0.47	0.50	0.55
Gender	Male	0.49	0.42	0.46	0.49	0.52	0.57
	Female	0.44	0.38	0.42	0.44	0.47	0.51
Age	18-44	0.47	0.40	0.43	0.47	0.50	0.56
	45-59	0.47	0.40	0.44	0.47	0.50	0.55
	60-79	0.46	0.38	0.42	0.46	0.49	0.54
	80+	0.43	0.35	0.40	0.43	0.47	0.52
Urban/Rural	Urban	0.47	0.40	0.44	0.47	0.51	0.56
	Rural	0.46	0.39	0.43	0.46	0.49	0.55
Region	North	0.48	0.41	0.45	0.48	0.51	0.56
	East	0.47	0.39	0.43	0.47	0.50	0.56
	South	0.45	0.38	0.42	0.45	0.48	0.53
	Northwest	0.47	0.40	0.44	0.47	0.50	0.55
	Northeast	0.48	0.41	0.45	0.48	0.51	0.57
	Southwest	0.45	0.38	0.42	0.45	0.48	0.53

Source: 2013 CEERHAPS-A.

Table 10-7 Recommended values of the surface area of body parts—feet

Category		Feet (m²)					
		Mean	P5	P25	P50	P75	P95
Total		0.11	0.09	0.10	0.11	0.11	0.12
Gender	Male	0.11	0.09	0.10	0.11	0.12	0.13
	Female	0.10	0.09	0.09	0.10	0.11	0.12
Age	18-44	0.11	0.09	0.10	0.11	0.11	0.13
	45-59	0.11	0.09	0.10	0.11	0.11	0.12
	60-79	0.10	0.08	0.10	0.10	0.11	0.12
	80+	0.10	0.08	0.09	0.10	0.10	0.12
Urban/Rural	Urban	0.11	0.09	0.10	0.11	0.11	0.13
	Rural	0.10	0.09	0.10	0.10	0.11	0.12
Region	North	0.11	0.09	0.10	0.11	0.12	0.13
	East	0.11	0.09	0.10	0.11	0.11	0.12
	South	0.10	0.09	0.09	0.10	0.11	0.12
	Northwest	0.11	0.09	0.10	0.11	0.11	0.12
	Northeast	0.11	0.09	0.10	0.11	0.12	0.13
	Southwest	0.10	0.09	0.09	0.10	0.11	0.12

Source: 2013 CEERHAPS-A.

Table 10-8 Recommended values of total body surface area and the surface area of body parts by province

Province/Municipality/ Autonomous region	Surface area * (m²)						
	Body	Head	Trunk	Arms	Hands	Legs	Feet
Total	1.6	0.12	0.60	0.24	0.08	0.47	0.11
Beijing	1.7	0.13	0.64	0.25	0.08	0.49	0.11
Tianjin	1.7	0.13	0.63	0.25	0.08	0.49	0.11
Hebei	1.7	0.13	0.63	0.25	0.08	0.48	0.11
Shanxi	1.7	0.13	0.62	0.25	0.08	0.48	0.11
Neimenggu	1.7	0.13	0.63	0.25	0.08	0.49	0.11
Liaoning	1.7	0.13	0.63	0.25	0.08	0.49	0.11
Jilin	1.7	0.13	0.62	0.25	0.08	0.48	0.11
Heilongjiang	1.7	0.13	0.62	0.25	0.08	0.48	0.11
Shanghai	1.7	0.13	0.62	0.24	0.08	0.48	0.11
Jiangsu	1.6	0.12	0.61	0.24	0.08	0.47	0.11
Zhejiang	1.6	0.12	0.60	0.24	0.08	0.46	0.10
Anhui	1.6	0.12	0.60	0.24	0.08	0.47	0.10
Fujian	1.6	0.12	0.59	0.23	0.08	0.46	0.10
Jiangxi	1.6	0.12	0.58	0.23	0.07	0.45	0.10
Shandong	1.7	0.13	0.63	0.25	0.08	0.49	0.11
Henan	1.7	0.13	0.62	0.24	0.08	0.48	0.11
Hubei	1.6	0.12	0.61	0.24	0.08	0.47	0.11
Hunan	1.6	0.12	0.59	0.23	0.08	0.45	0.10
Guangdong	1.6	0.12	0.58	0.23	0.07	0.45	0.10
Guangxi	1.5	0.12	0.57	0.23	0.07	0.44	0.10
Hainan	1.5	0.12	0.57	0.23	0.07	0.44	0.10
Chongqing	1.5	0.12	0.58	0.23	0.07	0.45	0.10
Sichuan	1.6	0.12	0.59	0.23	0.08	0.45	0.10
Guizhou	1.5	0.12	0.58	0.23	0.07	0.45	0.10
Yunnan	1.6	0.12	0.58	0.23	0.07	0.45	0.10
Xizang	1.6	0.12	0.58	0.23	0.07	0.45	0.10
Shaanxi	1.6	0.12	0.59	0.24	0.08	0.46	0.10
Gansu	1.6	0.12	0.61	0.24	0.08	0.47	0.11
Qinghai	1.6	0.12	0.61	0.24	0.08	0.47	0.11
Ningxia	1.7	0.13	0.62	0.25	0.08	0.48	0.11
Xinjiang	1.6	0.12	0.61	0.24	0.08	0.47	0.11

Source: 2013 CEERHAPS-A.
* Median values.

(Xiuge Zhao, Delong Fan, Nan Huang, Beibei Wang,

Xiaoli Duan)

11 LIFE EXPECTANCY

11.1 Introduction

Life expectancy is the number of years of life expected for individuals in a population, which should be considered when calculating the average daily dose during a health risk assessment.

11.2 Recommended values

The recommendation of life expectancy for the Chinese population is based on the life expectancy at birth for 2010 from the *China statistical yearbook 2012* (National Bureau of Statistics of the People's Republic of China, 2012). Table 11-1 provides the recommended values of life expectancy by province.

Table 11-1 Recommended values of life expectancy

Province/Municipality/Autonomous region	Life expectancy (years)		
	Combined	Male	Female
Total	74.83	72.38	77.37

Continued

Province/Municipality/Autonomous region	Life expectancy (years)		
	Combined	Male	Female
Beijing	80.18	78.28	82.21
Tianjin	78.89	77.42	80.48
Hebei	74.97	72.70	77.47
Shanxi	74.92	72.87	77.28
Neimenggu	74.44	72.04	77.27
Liaoning	76.38	74.12	78.86
Jilin	76.18	74.12	78.44
Heilongjiang	75.98	73.52	78.81
Shanghai	80.26	78.20	82.44
Jiangsu	76.63	74.60	78.81
Zhejiang	77.73	75.58	80.21
Anhui	75.08	72.65	77.84
Fujian	75.76	73.27	78.64
Jiangxi	74.33	71.94	77.06
Shandong	76.46	74.05	79.06
Henan	74.57	71.84	77.59
Hubei	74.87	72.68	77.35
Hunan	74.70	72.28	77.48
Guangdong	76.49	74.00	79.37
Guangxi	75.11	71.77	79.05
Hainan	76.30	73.20	80.01
Chongqing	75.70	73.16	78.60
Sichuan	74.75	72.25	77.59
Guizhou	71.10	68.43	74.11
Yunnan	69.54	67.06	72.43
Xizang	68.17	66.33	70.07
Shaanxi	74.68	72.84	76.74
Gansu	72.23	70.60	74.06
Qinghai	69.96	68.11	72.07
Ningxia	73.38	71.31	75.71
Xinjiang	72.35	70.30	74.86

Source: *China Statistical Yearbook 2012*, using life expectancy at birth for 2010.

(Yan Qian, Beibei Wang, Nan Huang, Suzhen Cao,

Xiuge Zhao, Xiaoli Duan)

12 RESIDENTIAL FACTORS

12.1 Introduction

Residential factors include the floor area, duration of heating, and ventilation rate, which are related to the space and air mobility of the indoor environment, and can affect the quality of indoor air. Floor area refers to the indoor area of a residential building, and the duration of heating is provided as the accumulated heating time in one year. The ventilation rate presents the longest time keeping window open for a household in the bedroom, study, and living room, in units of min/d.

12.2 Recommended values

The data for residential factors are from the 2013 *Environmental exposure related activity patterns research for the Chinese population (Adults)* (CEERHAPS-A). Tables 12-1 to 12-3 provide the mean and 5th-95th percentile data for floor area, heating duration, and ventilation time by gender, age, urban/rural location, and region. Table 12-4 provides the recommended values of residential factors by province.

Table 12-1 Recommended values of floor area

Category		Floor area* (m²)					
		Mean	P5	P25	P50	P75	P95
Total		126	40	76	100	150	300
Urban/Rural	Urban	120	40	70	92	140	300
	Rural	131	50	80	106	150	300
Region	North	107	40	66	90	120	220
	East	143	44	80	102	182	350
	South	135	40	80	120	165	300
	Northwest	108	50	79	100	120	200
	Northeast	80	40	60	80	90	120
	Southwest	143	60	90	120	180	300

Source: 2013 CEERHAPS-A.
* The floor area excludes open spaces, such as balconies and courtyards, as well as rarely used indoor areas (e.g., storerooms and basements).

Table 12-2 Recommended values of heating duration

Category		Proportion of population with heating activity (%)	Heating duration* (d/year)					
			Mean	P5	P25	P50	P75	P95
Total		74.2	105	28	60	110	150	180
Urban/Rural	Urban	74.5	104	20	60	120	150	180
	Rural	74.0	106	30	83	105	135	180
Region	North	88.7	120	60	90	120	150	180
	East	55.6	73	15	30	67	100	140
	South	56.6	78	15	50	88	120	150
	Northwest	99.5	142	90	120	150	180	190
	Northeast	99.3	147	90	120	150	180	180
	Southwest	66.2	90	22	60	90	120	180

Source: 2013 CEERHAPS-A.
* Accumulated heating time in 1 year for residents with heating activity.

Table 12-3 Recommended values of ventilation rate

Category		Ventilation rate* (min/d)					
		Mean	P5	P25	P50	P75	P95
Total		552	120	270	465	720	1,440
Urban/Rural	Urban	562	120	255	465	750	1,440
	Rural	544	135	270	453	720	1,320

Continued

Category		Ventilation rate * (min/d)					
		Mean	P5	P25	P50	P75	P95
Region	North	402	105	215	345	555	840
	East	602	135	300	510	810	1,440
	South	757	225	464	660	1,080	1,440
	Northwest	355	80	183	270	480	840
	Northeast	305	110	210	293	390	570
	Southwest	622	122	300	510	870	1,440
Season	Spring/Autumn	527	60	240	420	720	1,440
	Summer	840	180	480	720	1,440	1,440
	Winter	313	0	60	180	480	1,440

Source: 2013 CEERHAPS-A.
* The longest average ventilation time in the bedroom, study, and living room was reported as the ventilation rate for the household.

Table 12-4 Recommended values of residential factors by province*

Province/ Municipality/ Autonomous region	Floor area (m²)	Proportion of population using heating (%)	Heating duration** (d/year)	Ventilation rate (min/d)			
				Annual average	Spring/ Autumn	Summer	Winter
Total	100	74.2	110	465	420	720	180
Beijing	75	99.0	120	480	360	1,200	120
Tianjin	80	98.9	120	348	240	720	60
Hebei	90	99.6	150	300	240	600	60
Shanxi	80	99.7	135	240	240	480	60
Neimenggu	80	99.9	180	218	150	480	30
Liaoning	80	97.9	120	315	240	600	30
Jilin	80	99.9	150	240	180	600	0
Heilongjiang	70	100.0	180	285	240	630	0
Shanghai	93	62.7	40	495	480	600	240
Jiangsu	140	45.0	45	390	360	540	180
Zhejiang	150	37.0	50	570	540	720	360
Anhui	113	48.0	30	435	360	720	180
Fujian	180	6.7	30	900	900	1,440	600
Jiangxi	120	79.8	60	1,020	1,080	1,440	480
Shandong	80	89.8	105	425	300	980	60
Henan	100	59.5	90	545	480	840	180
Hubei	120	53.3	40	510	480	720	300
Hunan	130	98.4	90	630	600	960	360
Guangdong	110	15.8	30	900	900	1,440	600
Guangxi	120	66.8	60	540	480	780	300
Hainan	88	1.5	120	1,230	1,440	1,440	900

Continued

Province/ Municipality/ Autonomous region	Floor area (m²)	Proportion of population using heating (%)	Heating duration (d/year)	Ventilation rate (min/d)			
				Annual average	Spring/ Autumn	Summer	Winter
Chongqing	130	48.3	80	720	720	960	480
Sichuan	120	57.5	60	525	480	720	360
Guizhou	110	99.7	120	405	300	600	120
Yunnan	120	44.5	90	480	480	600	360
Xizang	220	90.8	165	210	240	309	60
Shaanxi	120	98.9	120	360	300	600	120
Gansu	100	99.2	150	375	300	600	120
Qinghai	85	99.9	180	195	180	360	60
Ningxia	94	100.0	150	375	360	720	60
Xinjiang	97	100.0	140	228	240	420	30

Source: 2013 CEERHAPS-A.
* Median values.
** Users only.

(Xiuge Zhao, Beibei Wang, Suzhen Cao, Ting Dong,

Xiaoli Duan)

References

AIST Research Center for CRM. 2007. Japanese exposure factors handbook, national institute of advanced industrial science and technology. Available from: https://unit.aist.go.jp/riss/crm/exposurefactors/english_summary.html [2014-07-10].

Committee on Improving Risk Analysis Approaches. 2011. Advancing risk assessment (Internet). Available from: http://www.epa.gov/region9/science/seminars/2012/advancing-risk-assessment.pdf [2014-08-27].

Duan XL, Nie J, Wang ZS, et al. 2009. Human exposure factors in health risk assessment. J Environ Health, 26(4): 370-373.

ECJRC. 2012. Expofacts database (Internet). Available from: http://expofacts.jrc.ec.europa.eu/index.php?category=database&source=db_a&PHPSESSID=dp2egk5j1450vas7qgoadrri81 [2014-07-10].

GM Richardson and Stantec Consulting Ltd. 2013. Canadian exposure factors handbook. Toxicology Centre, University of Saskatchewan, Saskatoon, SK Canada, Available at: http://www.usask.ca/toxicology/docs/cef [2014-01-23].

HRPAC. 2008. Compilation of exposure factors. Health Risks and Policy Assessment Center of Taihoku Imperial University.

Jang JY, Jo SN, Kim S, et al. 2007. Korean exposure factors handbook. Ministry of Environment, Seoul, Korea.

MEP. 2011. The 12th Five-year plan for the environmental health work of national environmental protection (Internet). Available from: http://english.mep.gov.cn/Plans_Reports/12plan/201201/P020120110355818985016 [2014-01-23].

MEP. 2013a. Exposure factors handbook of Chinese population. Beijing: China Environmental Science Press.

MEP. 2013b. Report of environmental exposure-related human activity pattern of Chinese population. Beijing: China Environmental Science Press.

National Bureau of Statistics of the People's Republic of China. 2012. China statistical yearbook 2012. Beijing: China Statistics Press.

Office of Health Protection of the Australian Government Department of Health, 2012. Australian exposure factor guide (Internet). Available from: http://www.health.gov.au/internet/main/publishing.nsf/Content/A12B57E41EC9F326CA257BF0001F9E7

D/$File/doha-aefg-120910.pdf [2014-07-24].

Phillips L, Moya J. 2013. The evolution of EPA's exposure factors handbook and its future as an exposure assessment resource. Journal of Exposure Science and Environmental Epidemiology, 23(1):13-21.

Phillips L, Moya J. 2014. Exposure factors resources: contrasting EPA's exposure factors handbook with international sources. Journal of Exposure Science and Environmental Epidemiology, 24(3): 233-243.

USEPA. 1989. Risk assessment guidance for superfund. Volume I: Human health evaluation manual (Part A). EPA/540/1-89/002.

USEPA. 1992. Guidelines for exposure assessment. EPA/600/Z-92/001.

USEPA. 1997. Exposure factors handbook. Office of Research and Development, Washington.

USEPA. 2011. Exposure factors handbook: 2011 Edition. Washington, DC.

Wang DL. 2005. General report of China national nutrition and health survey 2002. Beijing: People's Medical Publishing House.

Acknowledgments

This highlights handbook was funded by the Ministry of Environmental Protection of P.R. China (MEP) (Contract No.: EH(2011)-07-01, EH(2012)-07-01, EH(2013)-09-01, 201109064).

The authors would like to acknowledge the Department of Science, Technology, and Standards of the Ministry of Environmental Protection (MEP) for their financial and technical support. In particular, we would like to thank Dr. Yue Wan for her valuable assistance throughout the project.

In addition, the staff of the national and local Centers for Disease Control and Prevention (CDC) and all the field interviewers are appreciated for their help with the national survey. Moreover, we are very grateful for the valuable advice and comments provided by our advisory committee.

Printed in the United States
By Bookmasters